Tuning the Snowflake Data Cloud

Optimizing Your Data Platform to Minimize Cost and Maximize Performance

Andrew Carruthers

Apress®

Tuning the Snowflake Data Cloud: Optimizing Your Data Platform to Minimize Cost and Maximize Performance

Andrew Carruthers
Birmingham, UK

ISBN-13 (pbk): 979-8-8688-0378-9　　　　ISBN-13 (electronic): 979-8-8688-0379-6
https://doi.org/10.1007/979-8-8688-0379-6

Managing Director, Apress Media LLC: Welmoed Spahr
Acquisitions Editor: Shaul Elson
Development Editor: Laura Berendson
Editorial Project Manager: Gryffin Winkler

Cover designed by eStudioCalamar
Cover image by Silke from Pixabay

Distributed to the book trade worldwide by Apress Media, LLC, 1 New York Plaza, New York, NY 10004, U.S.A. Phone 1-800-SPRINGER, fax (201) 348-4505, e-mail orders-ny@springer-sbm.com, or visit www.springeronline.com. Apress Media, LLC is a California LLC and the sole member (owner) is Springer Science + Business Media Finance Inc (SSBM Finance Inc). SSBM Finance Inc is a **Delaware** corporation.

For information on translations, please e-mail booktranslations@springernature.com; for reprint, paperback, or audio rights, please e-mail bookpermissions@springernature.com.

Apress titles may be purchased in bulk for academic, corporate, or promotional use. eBook versions and licenses are also available for most titles. For more information, reference our Print and eBook Bulk Sales web page at http://www.apress.com/bulk-sales.

Any source code or other supplementary material referenced by the author in this book is available to readers on GitHub (https://github.com/Apress). For more detailed information, please visit https://www.apress.com/gp/services/source-code.

If disposing of this product, please recycle the paper.

For Diane, Esther, Josh, Verity, Evan, Violet, Jordan, and Beth

Table of Contents

About the Author

 **Andrew Carruthers** is the director for Snowflake distribution at the London Stock Exchange Group (LSEG). In this role, Andrew delivers several Snowflake accounts supporting Refinitiv "final mile" data product content delivery via Snowflake Marketplace, Private Listings, and Data Shares. He leads their Center For Enablement (C4E) in developing tooling, best practices, and training.

Previously, Andrew was responsible for the Snowflake Corporate Data Cloud at LSEG, which comprises two Snowflake accounts supporting an ingestion data lake and a consumption analytics hub and services a growing customer base of more than 7,000 end users. He also developed the Snowflake Landing Zone for provisioning Snowflake accounts conforming to both internal standards and best practices.

Andrew has more than 30 years of hands-on relational database design, development, and implementation experience starting with Oracle in 1993. Before joining the London Stock Exchange Group, he operated as an independent IT consultant, predominantly with major European financial institutions. Andrew is considered a visionary and thought leader within his domain, with a tight focus on delivery. Successfully bridging the gap between Snowflake technological capability and business usage of technology, he often develops proofs of concepts to showcase benefits leading to successful business outcomes.

Since 2020 Andrew has immersed himself in Snowflake and is considered a subject-matter expert. He is CorePro certified, contributes to online forums, and speaks at Snowflake events on behalf of LSEG. In recognition of his contribution to implementing Snowflake at LSEG, Andrew received the Snowflake Data Driver award, which recognizes a technology trailblazer who has pioneered the use of the data cloud within their organization.

Andrew has two daughters, both of whom are elite figure skaters. He has a passion for Jaguar cars, having designed and implemented modifications for them, and has published articles for Jaguar Enthusiast and Jaguar Driver. Andrew enjoys 3D printing and has a mechanical engineering workshop with a lathe, milling machine, and TIG welder, to name but a few tools, and enjoys developing his workshop skills.

About the Technical Reviewer

 **Nadir Doctor** is a database and data warehousing architect and a DBA who has worked in various industries with multiple OLTP and OLAP technologies. He has also worked on primary data platforms, including Snowflake, Databricks, CockroachDB, DataStax, Cassandra, ScyllaDB, Redis, MS SQL Server, Oracle, Db2 Cloud, AWS, Azure, and GCP. His major focus is health-check scripting for security, high availability, performance optimization, cost reduction, and operational excellence. He has presented at several technical conference events, is active in user group participation, and can be reached on LinkedIn.

Thank you to Andrew and all the staff at Springer. I'm grateful for the immense support of my loving wife, children, and family during the technical review of this book. I hope that you all find the content enjoyable, inspiring, and useful.

—Nadir

Acknowledgments

Thanks to the Apress team for the opportunity to deliver this book. Specifically, to Nirmal Selvaraj, Shaul Elson, and Mark Powers: thank you for your patient guidance, help, and assistance. Also, Nadir Doctor, thank you for delivering a comprehensive technical review. Your input provided more than just insight and valuable comments: I learned some new things, too. For those unknown to me, including editors, reviewers, and production staff, please take a bow. You are the unsung heroes who have made lightning strike three times (this is my third book with Apress).

To my very dear friend Andy McCann: I am more indebted to you than I can say. Your patience, insight, encouragement, and pragmatic approach provided much needed help and guidance. This book would not be anywhere near as complete or consistent without your input. I owe you more than a few beers.

To my friends at Snowflake who continue to both inspire and spur me on to bigger and better things: Jonathan Nicholson, Will Riley, Cillian Bane, James Hunt, Ben Conneely, and Adrian Randle. Keep on pressing forward; Snowflake is in good hands. Also, thanks to Jiaqi Yan and Minzhen Yang, whose inspiring talk sparked the idea that led to this third book in what has become a series. Little did I know back then just how hard this book would be to investigate, test, and write!

To John Ryan (`https://www.analytics.today/`), who put shape to my thoughts and inspired a section within this book, thank you.

To all my colleagues at London Stock Exchange Group (LSEG), specifically:

- **Corporate Data Cloud:** Nitin Rane, Srinivas Venkata, Matt Willis, Dhiraj Saxena, Bally Gill, Ramya Purushothaman, Radhakrishnan Leela, and Rajan Babu Selvanamasivayam

- **Snowflake Landing Zone:** Nareesh Komuravelly, Nathan Hawes, and Ravi Singh

- **Enterprise Data:** Kevin Whitchurch, Mike Frayne, Sahir Ahmed, Kalpesh Parekh, Matt Adams, Rajen Pather, R.Senthil Kumar, Prosenjit Chattoraj, and Chaitanya Kadiyam

ACKNOWLEDGMENTS

Take a bow. Thank you for your confidence, contribution, support, and help delivering world-class data products into LSEG distribution venues.

To my very dear friends Marco Costella, Martin Cole, Mike Sutherland, Lavkumaar Pandey, and Steve Loosley: keep on doing what you do best. If it isn't broken, don't fix it.

To my family, Esther and Josh; Verity, Evan, and baby Violet; and also Jordan and Beth: thank you. And to my wonderful girlfriend Diane, who continues her Snowflake journey: your smile brightens my day, and your presence makes me whole.

Will there be a fourth book in the series? Possibly. For now, it's time to rest and recharge. Eight months of preparation went into this book. I am not committing to writing a fourth book about Snowflake, though I do have enough material for half a book along with a title. And who knows what will happen after Snowflake Summit 2024?

CHAPTER 1

Tuning the Snowflake Data Cloud

This book continues from where both *Building the Snowflake Data Cloud* (Apress, 2022) and *Maturing the Snowflake Data Cloud* (Apress, 2023) left off. In this new volume, I deep dive into tuning Snowflake queries to deliver blisteringly fast performance along with a concurrent focus on cost-reduction efforts.

I unpack the core principles of how to approach performance optimization from several perspectives.

- Developers migrating existing applications to Snowflake must understand the pitfalls and "gotchas" that await the unwary.

- Cost management in an on-demand environment is a perpetual challenge, and squeezing every drop of performance from Snowflake is imperative.

- Optimizing warehouse size can reduce costs and improve throughput but often treats the symptoms and not the root cause of performance issues.

- Reducing micro-partition churn also reduces both storage and replication costs with the further benefit of reducing propagated data set latency, and I show you how.

- Remediating performance issues and refactoring production code to optimize performance involves trade-offs; there are no silver bullets!

- Updating existing Snowflake implementations to take advantage of new techniques is dependent upon understanding emerging product capabilities.

© Andrew Carruthers 2024
A. Carruthers, *Tuning the Snowflake Data Cloud*, https://doi.org/10.1007/979-8-8688-0379-6_1

In this book you will learn to develop tools and techniques based upon sound, proven, real-life scenarios. I use these tools and techniques daily, and as you become familiar with them, I hope you will too.

Performance tuning needs to be a continual activity. Data profiles change over time, and INSERT, UPDATE, and DELETE operations can cause skewed data where the distribution of data within a table or database becomes increasingly imbalanced or uneven. The impact of data skew over time can be significant, particularly when it comes to query performance.

All the examples used within this book were developed using a Snowflake trial account available at www.snowflake.com. Click the Start For Free button, and enter a few details to start a 30-day free trial account.

For those operating within a corporate environment, select Business Critical Edition because it is most likely the version used by your organization.

All the code samples in this book have been tested using Business Critical Edition and are believed to work with lower editions. You can find further details on Snowflake editions at https://docs.snowflake.com/en/user-guide/intro-editions.

I also assume you are familiar with the Snowflake user interface SnowSight (though the examples should work using SnowSQL or Visual Studio configured for Snowflake). You can find further details on SnowSight at https://docs.snowflake.com/en/user-guide/ui-snowsight. And for those starting their Snowflake journey for the very first time, start here: https://docs.snowflake.com/en/user-guide-getting-started.

I have attempted to divide this book content into readily consumed thematic chapters, and for the curious, the last chapter of this book on "gotchas" summarizes best practices. Before you jump straight to the end of this book, though, please read the intervening chapters as they will give you helpful context.

Last but certainly not least, you can find the Snowflake documentation at https://docs.snowflake.com/en/. Reading this book will definitely improve your learning curve; however, there are times where there is no substitute for reading official documentation (which is actually rather good); I will highlight some of it later, but for now, at least you know where it is.

Setting the Scene

I began writing this book in July 2023, a week after Snowflake Summit ended. My head was full of ideas, buzzing with the prospect of writing this book to impart my perspective and available wisdom on performance tuning Snowflake to a wider audience. What struck me was that, in just four years, Snowflake had transitioned from the cloud data warehouse of choice to a much richer and hard-to-define platform encompassing a wide variety of tooling, data formats, and capabilities.

Within this book I do not dive into the ever-expanding Snowflake product capabilities, instead preferring to focus on what some describe as the "black art" of performance tuning. By now, plenty of organizations have both ported applications to Snowflake and/or developed applications on Snowflake from scratch. The time is right for a book on Snowflake performance tuning to extract maximum value from these investments.

It would be too easy to cover what has already been described at an overview level by many vendors, some of whom are offering solutions that treat the symptoms and not the root cause. Conversations supported by Microsoft PowerPoint is one thing; practical techniques supported by hard and fast empirical evidence is entirely another. I prefer to demonstrate pragmatic approaches to resolving performance issues while developing tools to both educate and deliver a firm foundation for you to later build upon.

Snowflake is designed from the ground up to deliver optimal query performance with minimal user intervention. The "out-of-the-box" developer and user experience is truly exceptional, delivering astounding results for both data warehousing applications and, increasingly, much wider use cases including AI/ML applications.

In contrast to a provision-based model where you are constrained by your deployed infrastructure, Snowflake implements a consumption-based model: you pay for what you consume. Typically, provision-based infrastructure is idle for an average of 70 percent to 80 percent of the time, with occasional activity or, more commonly, overloaded activity peaks. In contrast, consumption-based models scale according to demand, providing performance elastically.

But this flexibility comes at a price: scalability and performance cost real money. you must therefore reconsider your approach in a consumption-based model and focus on reducing cost wherever possible. Costs are incurred when you execute code where you consume CPU and memory. In Snowflake parlance, CPU and memory are encapsulated within warehouses. You also incur costs for storage on a per-terabyte basis. At the time of writing, this is a direct pass-through cost from your cloud service provider (CSP). You also incur costs when you replicate data across regions and when you egress data from one CSP to another external location.

Unlike legacy products, Snowflake provides few levers and switches to influence system behavior and application performance, instead preferring to hide complexity to enable developers to focus on delivering business benefit. You might be lulled into a false sense of security by the ease with which you can port your applications into Snowflake, but this can be an expensive mistake.

Tuning the Snowflake Data Cloud is a project-oriented book with a hands-on approach to identifying migration and performance issues with experience drawn from real-world examples. As you work through the examples, you will develop the skill, knowledge, and deep understanding of Snowflake tuning options and capabilities while preparing for later Snowflake features as they become available. Your Snowflake platform will cost less to run and will improve your customer experience.

It is important to note that Snowflake is a constantly evolving product, and therefore best practices will change over time. You should not expect the advice, hints, and tips in this book to be static; this book offers what I know right now, with both eyes on the future.

Regardless of your relational database management system (RDBMS) experience, it's safe to say some of your performance tuning skill, knowledge, and expertise is directly transferable. Equally, some prior learning is not transferable; a degree of unlearning will be required, and for those working on both legacy RDBMS and Snowflake, the operating paradigms are distinctly different.

I next discuss some common themes.

Use Cases for Snowflake

Fundamentally, the underlying CSP storage (whether S3, Azure Blob, or Google Cloud storage) and Snowflake's immutable storage policy dictate the supported transaction style, with data warehousing preferred over online transaction processing (OLTP).

As a general rule of thumb, Snowflake prefers high-volume bulk-load operations supporting analytics workloads. Low-latency, high-volume transactions are not yet common workloads for Snowflake.

The forthcoming Unistore workload joining transactional and analytical data via hybrid tables may change this perception.

Hybrid tables are not yet generally available.

You can find further details on Unistore at `https://www.snowflake.com/en/data-cloud/workloads/unistore/`.

Rapid data ingest options via data streaming requiring low latency for low-data volume is another common use case. I recommend the "Tour of Ingest" at `https://quickstarts.snowflake.com/guide/tour_of_ingest/index.html`.

For those looking to understand a much wider suite of Snowflake use cases, please investigate all the various quick starts at `https://quickstarts.snowflake.com/`.

Provision or Consumption Model

Performance tuning in a provision-based model has fixed constraints; you cannot simply pop down to the data center and plug in more memory or replace your hardware with faster devices. Without preplanned system downtime for upgrades along with the service disruption caused, you are limited to eking out every small performance increment from your existing hardware using any and all levers provided by your operating system vendor, RDBMS vendor, network tooling, and storage vendor. And all of these require deep subject-matter experts (SMEs) in each topic to interact and define optimal patterns for repeatability. Well, that's the intent, but as you all know, reality does not always match expectations.

In sharp contrast, a consumption-based model such as Snowflake removes many historically familiar tuning options and levers; no longer are you able to tune the operating system and change the RDBMS kernel settings. Instead, Snowflake implements a managed service where you pay for what you consume, and this brings about totally different challenges. Gone are the provision-based constraints, but leaving aside the shift to a security focus, which a consumption-based model requires, you replace the provision-based hardware constraints with two new major challenges: cost and performance optimizations.

There is one crucial but often overlooked benefit to adopting a consumption-based model. Snowflake performance has steadily improved since reported performance metrics were first established in August 2022, for two reasons.

- Optimizer performance has steadily been enhanced over time, realizing tangible benefit to overall query execution times.

- CSP hardware replacement programs for obsolete or end-of-life hardware utilize the latest hardware automatically providing performance uplifts.

In August 2022, Snowflake began to record these zero-cost performance benefits. Figure 1-1 illustrates the Snowflake Performance Index, which can be found at `https://www.snowflake.com/en/data-cloud/pricing/performance-index/`.

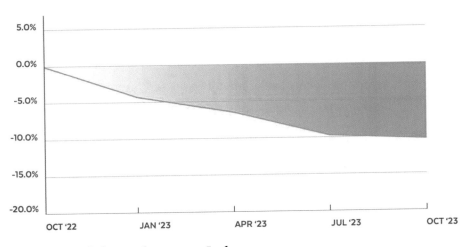

Figure 1-1. *Snowflake Performance Index*

The trend is set to continue as Snowflake is committed to improving its code base and CSPs periodically replace hardware due to their preventative maintenance policies.

The key takeaway is to periodically monitor your system performance for improvement or degradation over time and take into consideration the probability of Snowflake and CSP changes positively affecting your consumption costs.

Still with me? Good, let's explore common Snowflake starting points (although these are not exhaustive, and your steps may differ).

Refactor or Redesign

Refactoring is the process by which you simplify an existing code base while retaining the original functionality. You might choose to refactor code to take advantage of new performance enhancements, implementing both common design patterns and code structures while improving the overall implementation. Regardless of the rationale for refactoring, the aim is to preserve the original functionality; there should be no discernable behavior differences from the original. Thus, retesting should be as simple as re-running the original test cases utilizing the same inputs.

Refactoring is not intended to address software flaws. It is perfectly valid (and desirable) when refactoring code to improve performance and scalability while preserving the original functionality.

In contrast, redesign may not preserve the original functionality and often modifies, extends, or otherwise improves the functional utility of the component in accordance with the design specification.

Redesign is intended to address software flaws. It is perfectly valid (and desirable) while redesigning code to improve performance and scalability.

Within this book I will use the previous definitions; however, as you will see later, sometimes the boundaries are blurred.

Application Migration to Snowflake

Migration from legacy RDBMS to Snowflake is a common driver to unleash huge performance benefits while moving to CSP infrastructure. I do not discuss in detail "how" to migrate applications to Snowflake nor leverage CSP infrastructure within this chapter, but note these steps are typically performed:

- **Planning:** Developing a project plan incorporating scope, funding, resources, and timeline.

- **Code conversion:** Writing SQL statements, Data Definition Language (DDL), user-defined functions, stored procedures syntax, language conversion.

- **Entitlements:** Refactoring the legacy application security model to use a Snowflake role-based access model (RBAC) model.

- **Data migration:** Porting the application data into Snowflake and establishing ingestion pipelines and processes.

- **Data consumption:** Re-engineering the application outbound data consumption processes.

- **Platform security:** Adding security. I cover this point in great detail in the *Maturing the Snowflake Data Cloud* book.

- **Performance:** Optimizing Snowflake is the core subject matter of this book.

- **Testing:** Perform back-to-back testing to ensure equivalent outputs for known inputs are delivered, along with the all-important and expected performance benefits.

- **Documentation:** No migration activity is complete without exhaustive documentation.

The most time-consuming and difficult step to determine is code conversion; no two applications have the same profile or migration objectives. Migrating an application for archive legacy purposes to retain data for a specific period will be very different from migrating an active, in-use application.

Refactoring code is expensive, and finding empirical metrics is hard. As a rough guideline, you can expect refactoring costs to be at least four times the cost of developing code from scratch. This guesstimate includes understanding the original code; you can substitute your own multiplication factor taking into consideration the availability of experienced resources and detailed documentation.

Another considerable challenge is ensuring your migrated application functionality matches the source application. I call this out as source applications are not typically fixed in time; enhancements and bug fixes cause divergence that must be considered when porting to Snowflake.

Migration Guides

Snowflake offers a number of legacy RDBMS guides to help you port applications to Snowflake. Some of these are listed here and may require your contact information before access is enabled:

- `https://www.snowflake.com/wp-content/uploads/2020/05/oracle-to-snowflake-technical-migration-guide.pdf`

- `https://www.snowflake.com/resource/microsoft-sql-server-to-snowflake-migration-reference-manual/`

- `https://www.snowflake.com/wp-content/uploads/2020/08/teradata-to-snowflake-migration-guide.pdf`

- `https://www.snowflake.com/resource/spark-to-snowflake-migration-guide/`

Aside from these product specific listings, other migration guides and additional related information are available at `https://www.snowflake.com/en/resources/?tags=content-type%2Fmigration-guide&searchTerm=migration`.

Migration Options

In this section I will identify some options to migrate applications to Snowflake and later focus on performance and cost optimization.

Character set conversions require special attention outside of the scope of this book.

SnowConvert

In January 2023 Snowflake acquired SnowConvert from Mobilize.net, a toolkit for migrating customer workloads from legacy RDBMS to Snowflake. SnowConvert automates schema and functional component conversion to Snowflake from a variety of legacy RDBMSs.

Since the Snowflake acquisition, SnowConvert has become the Snowflake Professional Services (PS) tool of choice for application migration. Naturally, you do not have access to SnowConvert directly, but you can find further information at `https://www.mobilize.net/`.

Manual Schema Conversion

Depending upon your requirements and perceived application code complexity, it is possible to convert schema objects to Snowflake syntax relatively easily. One successfully used approach involves the use of the shell scripts awk and sed to refactor Data Definition Language to Snowflake syntax. Note this approach does not address performance tuning concerns but does provide a baseline from which to start.

Manual schema migrations are relatively straightforward; however, there are some caveats.

Identifying source character sets can be challenging. Sometimes character set corruption has occurred before data was ingested within the application to be ported; therefore, reconciliation when converted to the Snowflake default UTF-8 character set is impossible.

User-defined types must be reconciled back to their base data types, which in most scenarios will be the supertype rather than subtype. For example, declare FLOAT, DECIMAL, MONEY, NUMBER with or without precision, etc.

Some objects do not lend themselves to direct conversion; for example, this nonexhaustive list of Oracle to Snowflake migration challenges will require remediation:

- Snowflake does not support ROWID.

- Within tree walks, Snowflake does not explicitly support LEVEL.

- Complex materialized views are not directly supported; dynamic tables are an equivalent, but at the time of writing this feature is not generally available.

- Snowflake NULL treatment is ANSI compliant; Oracle NULL treatment is not.

- Embedded documents are often encoded, encrypted, or compressed using proprietary algorithms.

- Snowflake doesn't have synonyms and relies upon search_path.

Likewise, SQLServer to Snowflake migration challenges may be found by doing the following:

- Resolving user-defined types and platform-specific data types to their equivalent Snowflake supertypes

- Using SQL Server syntax that diverges from the ANSI standard

In general, across many legacy RDBMSs, you will also find these differences:

- Date functions, format specifiers, and time zones, in common with other legacy RDBMSs.

- Absence of index support in Snowflake for standard tables (though forthcoming Unistore hybrid tables do use indexes).

- Cluster key terminology and usage are not the same across legacy RDBMSs and Snowflake.

- Declared but not enforced constraints except for NOT NULL in Snowflake.

I offer these as a short and incomplete list to give you an idea of the differences between legacy RDBMSs and Snowflake.

The ACID tests are whether data correctly loads into the Snowflake objects migrated from the source and all regression tests run clean.

Functionality Lift and Shift

Assuming the schema has been ported to Snowflake, the task of porting functional components remains. Tools like SnowConvert claim to port stored procedures to JavaScript equivalents, and I have no evidence to suggest otherwise.

My concerns relate to the quality of SQL ported from source.

Do not expect unmodified SQL statements to be optimally performant in a Snowflake environment. Experience has proven that you must tune all SQL code for the platform and not assume everything will run "just fine" in Snowflake.

Typically, you will see a performance boost because of the massively parallel processing (MPP) capability Snowflake brings. You must not be complacent in lifting and shifting code and then accepting the new MPP performance benefit as evidence of success.

Instead, I suggest identifying the top 10 longest running queries and then optimizing their performance using the techniques outlined in this book. A note of caution: you are looking for repeating SQL statements; one-off data loads should be excluded.

My experience of tuning highly complex ported queries shows an upward of a 20 percent performance improvement, which directly translates to a cost reduction. Once the top 10 queries have been optimized, recheck and identify the next top 10 longest running queries and repeat performance tuning.

Greenfield Development

Several steps are the same as with application migration though with different boundaries.

- **Planning:** Develop a project plan incorporating scope, funding, resources, and timeline.

- **Entitlements:** Determine the application entitlement model using RBAC.

- **Data model:** Design and implement the application data model.

- **Data consumption:** Create the application outbound data consumption processes.

- **Platform security:** I cover this point in great detail in the *Maturing the Snowflake Data Cloud* book.

- **Performance:** How to optimize Snowflake is the core subject matter of this book.

- **Testing:** Perform testing to ensure all functional and nonfunctional requirements are met.

- **Documentation:** No development activity is complete without complete documentation.

I do not dwell on greenfield development as this is a well-trodden path with many skilled practitioners ready and willing to undertake Snowflake development.

Replication Considerations

A distinct advantage of porting applications to Snowflake is the ability to utilize Secure Direct Data Shares, Private Listings, and Snowflake Marketplace. As noted, you will also incur costs when you replicate data across regions and when you egress data from one CSP to another external location.

While data sharing and replication may not be used ubiquitously, for those who do use them, replication costs can far exceed the warehouse runtime costs to generate the original data sets. Optimizing data transfer will reduce replication costs; I show you how to do this later. For a taste of things to come, by redesigning your approach to storing data, it is possible to significantly reduce both storage and replication costs.

Tune the Design

Regardless of whether you migrate an application or are delivering a greenfield development, you must tune your design and adopt an approach that has the best chance of success. Various studies have shown that between 66 percent and 85 percent of all application deliveries fail. How you set out is a key determinant for success or failure.

Operating in a high-pressure, delivery-focused environment can lead you to ignore the importance of tuning your design. As your projects progress, the incremental cost of refactoring your design increases, so you must approach any new delivery with caution. Take the time to validate your approach and seek wisdom from those who have successfully implemented Snowflake applications, recognizing there are not many people who have done this.

Tuning Snowflake designs is dependent upon fully understanding the underlying Snowflake platform architecture. Sadly, there are plenty who understand enough to treat the symptoms but not enough to address root-cause issues. As my good friend Andy McCann says, "Good practice travels far," but note the pace of change of Snowflake delivery is accelerating, and what was considered a best practice a year ago may not stand the test of time now.

For those with deeper pockets and appetite, I strongly advocate that Snowflake PS is engaged at the earliest opportunity to validate and identify optimally performant patterns. You can find further information on Snowflake PS at `https://www.snowflake.com/snowflake-professional-services/`.

Snowflake training that may lead to certification will accelerate learning. Depending upon your current RDBMS knowledge and career aspirations, different courses will appeal. You can find further information at `http://learn.snowflake.com/en/`.

An alternative but slower route to success for those without a strong RDBMS background is via Snowflake University, where an introduction to Snowflake development course is available. This course is free and self-paced, and Snowflake trial accounts can be used. You can find further information at `https://www.snowflake.com/snowflake-essentials-training/`.

Otherwise, for those seasoned practitioners looking for inspiration, code samples, and walk-throughs, there is no better suite of resources than those found at `https://quickstarts.snowflake.com/`.

Ignore tuning your design at your peril; this step is the lowest cost while providing the biggest "bang per buck" regardless of platform. This advice will serve you well throughout your IT career.

Your First Optimization

You declare compute in T-shirt-sized warehouses according to the perceived demand your SQL statement will place upon CSP resources.

Every time you execute a SQL statement requiring a warehouse, you incur cost. By default, Snowflake delivers a single warehouse called `compute_wh`, and your first optimization must be to ensure the default warehouse. Every other defined warehouse runs for the minimum time before suspending. The default setting for auto-suspending `compute_wh` is 10 minutes, or 600 seconds.

We use the `auto_suspend` attribute with a minimum of 60 seconds as shown next:

```
SHOW warehouses;
ALTER WAREHOUSE compute_wh SET auto_suspend = 60;
```

Every warehouse regardless of size runs for a minimum of 60 seconds, with per second billing thereafter.

Warehouse events tell us information relating to warehouses; we use this information primarily to check the RESUME and SUSPEND conditions where the next SQL statement can be used.

```
SELECT *
FROM   snowflake.account_usage.warehouse_events_history
WHERE  warehouse_name = 'COMPUTE_WH'
ORDER BY timestamp DESC;
```

Figure 1-2 shows sample output.

TIMESTAMP	WAREHOUSE_ID	WAREHOUSE_NAME	CLUSTER_NUMBER	EVENT_NAME
2023-07-09 00:19:22.722 -0700	1	COMPUTE_WH	0	SUSPEND_WAREHOUSE
2023-07-09 00:19:22.722 -0700	1	COMPUTE_WH	0	SUSPEND_CLUSTER
2023-07-09 00:19:22.674 -0700	1	COMPUTE_WH	0	SUSPEND_WAREHOUSE

Figure 1-2. *Warehouse events history*

In addition to determining warehouse runtime, you can also see CLUSTER_NUMBER indicating scaling out; I discuss warehouse clustering later.

Optimally sizing warehouses and the clustering factor provides a significant opportunity to reduce costs, a theme I will return to later.

Optimizer Approach

Many of us remember a legacy "rules-based" approach to defining an efficient execution plan, a suite of predefined rules applied to the SQL statement used to derive the optimal execution path. The "rules-based" approach did not require statistics; if the rules quality and coverage did not cater for the SQL statement being executed, then poor performance would most likely ensue.

For most if not all RDBMSs, "rules-based" optimizers have largely been replaced by "cost-based" optimizers where real-world statistics inform sophisticated algorithms to evaluate and select execution plans. Cost-based optimizers generally outperform rule-based approaches in terms of query optimization effectiveness and adaptability.

Snowflake has adopted a simplified approach to delivering their optimizer by only supplying a cost-based optimizer. Furthermore, for their optimizer, Snowflake by design leaves as little to the user as possible. For example, unlike some legacy RDBMS, you cannot add hints to influence the generated optimizer query plan.

Plan stability is essential for predictable repeat performance, and all RDBMS vendors strive to achieve this objective. Snowflake is no exception where the key objective is to create a robust cost-based optimizer delivering stable execution plans. Additionally, Snowflake focuses on optimizations for analytics workloads and support for all data models including third normal form, data vault, and star schema. You also know Snowflake has implemented many non-cost-based optimizations and continues to work on eliminating nonperformant edge cases too, all part of the continual product improvements.

Query Parsing Order

I have discussed DDL and now will move on to discussing Data Manipulation Language (DML). The most common DML statement you will encounter is SELECT, where you retrieve data from Snowflake. I often refer to a SELECT statement as a *query*.

To effectively tune queries, you must understand how SELECT statements are executed. The following code sample demonstrates a simple single table query syntax:

```
SELECT DISTINCT column
FROM    mytable
WHERE   constraint_expression
GROUP BY column
HAVING    constraint_expression
ORDER BY column ASC/DESC
LIMIT   count OFFSET start_point;
```

SQL is passed through several subsequent stages that I describe later, but for now, I'll focus on identifying the order in which each part of the SQL statement is parsed.

To reduce complexity, you will not consider inline functions, common table expressions (CTEs), tree walks, set operators, and other advanced constructs.

Figure 1-3 illustrates the order in which SQL operations are actioned.

Figure 1-3. *SQL order of operations*

The next section provides some detail on "how" SQL is parsed and offers some broad usage advice based upon real-world experience.

FROM Clause

The first step is to identify the object(s) where data is stored. The FROM clause identifies the objects, in your example, a single table. But you often join to additional tables, and the order in which you join the tables has significance to the Snowflake optimizer.

Always put the smallest table first when joining tables, noting tables may grow over time.

You might also see JOIN criteria to reference further tables, regardless of whether a single table or multiple tables. A lookup is performed to ensure specified named object(s) exist within the metadata repository. The FROM and JOIN clauses provide the total accessible data set for the query referencing the attributes (table columns) and rows (the actual data).

I prefer to fully qualify object names using database.schema.object notation to prevent ambiguity when referencing source objects.

WHERE Clause

The FROM clause identifies the full scope of accessible objects and attributes for the query. The WHERE clause—also known as the *predicate*—identifies how the in-scope objects relate to each other, also defining filters or constraints.

You might see multiple predicates applied using AND / OR syntax; each is a filter or constraint to the returned data set.

Predicates implement join conditions between tables and are essential for developing optimally performant code. Where two or more objects are accessed and not joined, a Cartesian product (also called a *join explosion*) results.

Failure to join tables in a WHERE clause results in a Cartesian product.

When generating high volumes of test data, you might deliberately choose to omit a join condition. This is a rare but acceptable use case noting the performance implications.

GROUP BY Clause

When aggregating attributes, you add a GROUP BY clause. Grouping ensures the returned data set contains rows equal to the unique values in that column.

This optional clause is used for aggregations only.

GROUP BY is used only with aggregate queries.

HAVING Clause

A HAVING clause is permissible only in conjunction with a GROUP BY clause, to filter results. A common use case is to filter by a specific count to identify duplicate records.

This optional clause is used for filtering aggregated data.

HAVING is used only with aggregate queries.

SELECT Clause

With your subset of objects, attributes, and data identified, any single row or aggregate expressions within the SELECT statement are computed.

The optimal number of attributes for SELECT statements is 10 or fewer. This figure was disclosed during a discussion with Snowflake staff. Wide tables with many rows do not perform particularly well, and the use of temporary tables was also suggested as an appropriate intermediate step.

Through hands-on experience, you found SELECT * performance to be suboptimal across a wide range of queries migrated from legacy RDBMS. I therefore strongly recommend explicit attribute references.

Always refactor SELECT * to reference explicit attribute names.

One exception is that I have found SELECT * with an appropriate LIMIT acceptable when sampling data as a pre-cursor to assist the development process.

As with everything performance related, test, test, and test again. Don't rely upon what you read either here or elsewhere, but instead prove it empirically. Your experience using real-world conditions is what matters.

DISTINCT Clause

Where only unique rows are required to be returned, a DISTINCT clause may be used.

This optional clause is used to return unique rows only.

DISTINCT forces an aggregation operation visible within the query profile. I discuss this later, but for now it is sufficient to know there is an impact for using a DISTINCT clause. I find SELECT DISTINCT acceptable when sampling data as a pre-cursor to assist the development process.

The use of DISTINCT in production code often indicates a missing join condition within a query, an incomplete/incorrect database design, or an expedient solution being used to resolve a data quality issue. Regardless of the root cause, I recommend always investigating such occurrences.

ORDER BY Clause

For a variety of reasons you may need to order returned data sets in either ascending or descending order. For example, in a graphical user interface, you may want to display search results in alphabetical order.

This optional clause is used to sort returned data sets.

Ordering returned data sets forces a sort operation visible within the Query Profile. I discuss this later, but for now it is sufficient to know there is an impact for using an ORDER BY clause.

LIMIT/OFFSET

Sometimes you will need to sample data, and LIMIT provides the operator to restrict the returned data set to a known sample size. Likewise, you might want to start your sample from a nominal position within the returned data set, and OFFSET provides this capability.

This optional clause is used to return a subset of the returned data set.

Using LIMIT/OFFSET can be useful while testing code but generally is not used in production code; the local code comments will no doubt self-document to explain "why."

SQL Joins

You can join tables within the WHERE clause by declaring a relationship between attributes in both tables, preferably by using primary keys and foreign keys but alternatively by using natural keys too. The syntax for expressing join conditions comes in two forms, discussed next.

Explicit Join Notation

Explicit join notation is considered a best practice and should be considered for inclusion within SQL coding standards. As the name suggests, the join type for each object accessed is explicitly stated.

You can find further information on SQL at https://en.wikipedia.org/wiki/Join_(SQL).

Here's an example query using explicit join notation:

```
SELECT count(1)
FROM   partsupp_baseline    ps
INNER JOIN part_baseline    p
   ON ps.ps_partkey         = p.p_partkey
INNER JOIN supplier_baseline s
   ON ps.ps_suppkey         = s.s_suppkey;
```

Implicit Join Notation

Considered by many to no longer be a best practice, this is my preference and widely used throughout both this book and my previous books.

I have found implicit join notation to offer these advantages over explicit join notation:

- Implicit join notation avoids forward declaration errors.

- To me, implicit join notation is cleaner and easier to read.

Here's the same query using implicit join notation:

```
SELECT  count(1)
FROM    partsupp_baseline ps,
        part_baseline     p,
        supplier_baseline s
WHERE   ps.ps_partkey     = p.p_partkey
AND     ps.ps_suppkey     = s.s_suppkey;
```

You will encounter this SQL statement in Chapter 3.

Forward Declaration Errors

A forward declaration error is a parsing failure and occurs when a reference is made to an object or attribute that has not yet been declared. From the earlier discussion on query parsing order, you know the FROM clause is parsed first.

- For explicit join notation, each joined table is evaluated *in declaration order* along with its joining criteria.

- For implicit join notation, all tables are evaluated *at the same time*; then all predicates are evaluated.

As a consequence of the evaluation order, it is possible to reference join keys that have not yet been parsed but are declared later in the SQL statement.

Valid arguments for mandating explicit join notation are to both reduce code errors and enforce discipline when accessing data models. However, readability is important too.

We leave you to decide which approach suits your requirements best.

Introspection Calls

In Snowflake, an introspection call is a SQL statement used to interrogate the account usage store or information schema of a particular database to identify metadata for objects, columns, and their attributes. Since Snowflake is typically operated in a highly controlled, secure manner and one significant use case is data warehousing, there is an expectation for object metadata to remain static. Best practice dictates both schema and object structural changes should be made only in accordance with a rigorous change control process. Introspection calls should therefore be able to rely upon a relatively static suite of metadata.

In practice, you have found some Snowflake metadata lookups run slower than expected when compared to an alternative RDBMS. The root cause appears to be unexpected or unpredictable changes made to object definitions causing ad hoc metadata changes. You might experience this phenomena where self-service has been implemented for end users to own, manage, and maintain their own schemas where unpredictable and ungoverned schema changes occur.

Further investigation is required as you cannot yet categorically identify the root cause of slow-running metadata queries. I suggest queries using the account usage store may be improved by making local copies of referenced views into tables (thanks to Nadir Doctor for this tip). There is further sparse anecdotal evidence of this phenomena that may be obtained using common Internet search engines.

Optimizer Statistics

Optimizer statistics are information and data values collected about the database objects used by the query optimizer to inform decisions on how to execute SQL queries efficiently. In contrast with some legacy RDBMSs, Snowflake guarantees the optimizer statistics are always up-to-date; there is no delay. And also unlike some legacy RDBMSs, for Snowflake there is no exposed capability to collect, delete, or manage statistics; this is by design.

Snowflake captures the following statistics:

- Table and micro-partition

    - Row count

    - Size in bytes (including compression information)

    - File reference

    - Table version

- Clustering
    - Total micro-partitions
    - Micro-partition overlap values
    - Micro-partition depth
- Column
    - Max/min value range
    - The number of distinct values
    - NULL count
- Subcolumn
    - Statistics for common paths in semi-structured data

Snowflake caches statistics within the Cloud Services layer. Statistics are used as inputs to the optimizer cost model and for micro-partition pruning; both are discussed later.

Summary

In this chapter, I laid out the rationale for this book while determining the core focus: our primary objective is to focus on cost optimization for all Snowflake activity, not just single query tuning.

I drew comparisons between provision-based models and consumption-based models, noting the focus shift to security, cost, and performance. Your decisions on how to approach application development and migration have material implications for cost optimizations. When replicating data sets, replication costs can far exceed the initial cost of generating the data sets.

I then covered how SQL is parsed, noting the statement parsing order and some potential issues you may encounter before discussing optimizer statistics.

Having established baseline information in preparation for looking deeper into performance tuning, it's time to move on to the next chapter.

CHAPTER 2

The Query Optimizer

Query optimizers reduce the cost of queries while retaining the original intended functionality. Furthermore, query optimization seeks to reduce the volume of data accessed, further reducing costs.

A lot happens within the Snowflake query optimizer, and not every detail is known to the wider user community. I have drawn upon available sources to piece together what can be shared, although the optimizer behavior might have changed by the time you read this as a natural consequence of development and maintenance. However, there is value in reading this chapter as I hope it will help shape your thinking when designing SQL statements.

Building upon the information presented in Chapter 1, in this chapter I discuss various aspects of the query optimizer beginning with the lifecycle of a query. I then move on to discussing what happens within the planner and optimizer, a fascinating subject in its own right.

The key message for this chapter is to adopt the KISS principle, better articulated here: `https://en.wikipedia.org/wiki/KISS_principle`. The same principle is equally well stated by Tony Robbins: "Complexity is the enemy of execution" (`https://www.youtube.com/watch?v=oOPweQFmJpI`). We will return to this theme frequently.

A guiding principle I use for determining the quality of code is to see how cleanly written and laid out each SQL statement appears. If the SQL looks good, is well formatted, and is readable, then it is most likely the developer has taken great care to ensure optimal execution performance. You should not lose sight of how long it takes to refactor code; all good developers hate cleaning up other people's mess.

For those just starting out: help those who support your code by delivering high-quality, easily understood, and well-documented artifacts.

© Andrew Carruthers 2024

A. Carruthers, *Tuning the Snowflake Data Cloud*, https://doi.org/10.1007/979-8-8688-0379-6_2

No query optimizer is able to guess at the intended data set outcome. As the saying goes, garbage in, garbage out, and you are responsible for ensuring the quality of your submitted SQL statements.

Finally, I must pay tribute to both Jiaqi Yan, principal software engineer, and Minzhen Yang, principal engineer and tech lead, both at Snowflake, for their comprehensive explanation of the Snowflake query optimizer at Snowflake Summit 2023. I have derived some information for this chapter from their presentation as well as embellished it with my own understanding and knowledge. Any omissions, misrepresentation, or errors are mine alone.

Query Lifecycle

At a superficial level, we submit queries, and sometime later we receive results. Simple. Or is it? The book you are reading should automatically answer the question of simplicity. Ideally this book would not be necessary, because the developers try very hard indeed to remove complexity from Snowflake.

The absence of levers and switches to influence system behavior and application performance is a key indicator of how successful the developers have been. Understanding the query optimizer implementation unlocks pathways to delivering SQL.

Wherever possible, simplify your SQL; do not write convoluted or hard-to-follow code.

Figure 2-1 shows an overview of the query lifecycle, which is explained further in the corresponding bullet points.

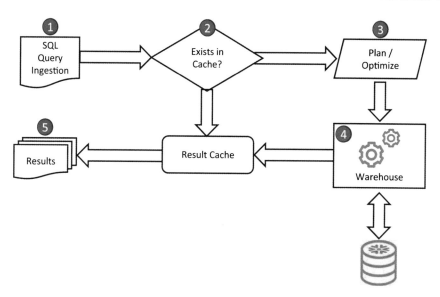

Figure 2-1. *Query lifecycle*

Query Overview

I will now explain at a high level the steps highlighted within Figure 2-1.

1. Snowflake ingests a SQL statement; you do not specify the source as there are many and varied inbound connection paths.

2. If the query exactly matches a previously run query *and* the data has not changed, then the result set is returned from the result cache.

3. Otherwise, the query planner and optimizer use metadata to work out the exact data set and lowest cost access path to satisfy the SQL query.

4. The warehouse identifies and retrieves the exact data from the local or remote disk and then returns the data to the cloud services.

5. The result set is returned to the client and stored in the result cache for reuse.

Figure 2-1 is just an overview; a more comprehensive explanation is provided shortly.

Naturally, queries may fail to execute for a variety of reasons, some of which I briefly discuss next.

Query Failure

Queries may fail to process for a variety of reasons. I cannot list all possible query execution failures though they largely fall into two categories: Queries that fail due to not passing through the query optimizer processing through to execution, and queries that successfully begin execution but subsequently fail due to infrastructure capacity or interconnectivity failure.

An example of infrastructure failure could be an occasional warehouse failure where Snowflake automatically recovers to complete the query execution transparently noting the result set may be slightly delayed while detection and recovery occurs.

The quality of our SQL statements can also lead to execution failure. As an example, a missing join condition often results in a Cartesian product (also known as a *cross-join*) where unexpectedly large result sets are generated resulting in spills or out-of-memory warehouse failures.

In my 30+ years of experience across a variety of RDBMS platforms, it's usually my code that is at fault.

Some examples of why queries may fail to execute include the following:

- Invalid syntax

- Inaccessible object

- Out of memory

- Client process failure

- Network or interconnect failure

- Warehouse failure

- Other unspecified reason

I will not dive into the root causes of each potential failure. The following information enables you to identify where some query failures occur along with sufficient context to explain "why" such failures may occur.

Query Compilation

In this section I explain how a noncached query is processed.

If the statement in step 2 of Figure 2-1 ("If the query exactly matches a previously run query AND the data has not changed, then the result set is returned from the Result Cache") is TRUE, then this section is not executed.

Figure 2-2 illustrates the processing path followed when step 2 of Figure 2-1 is FALSE, corresponding to step 3, which is "Plan/Optimize":

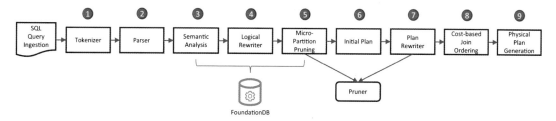

Figure 2-2. *Query compilation steps*

Before we deep dive into the Snowflake query optimizer, it is worth saying that all RDBMSs have a query optimizer. We therefore rely upon common, in-use terminology, and for those new to understanding query optimizers, we briefly explain terminology shortly.

Before we dive in, our aim is to impart sufficient information to cause you to think about how your code will execute once submitted to the query optimizer. The easiest, cheapest, and most effective performance tuning occurs before a single line of code is written, that is, during the design phase. All the information presented within this book is intended to provide you with the tools to tune your code before submission to the Snowflake query optimizer.

Also note not every stage or step within a stage is mandatory. The query optimizer may choose to skip stages or individual steps where appropriate. As an example, if no CTE is detected within the submitted SQL statement, there is no need to expand CTEs. Snowflake refers to this process as *automatic skipping of redundant stages*. I indicate where stages and steps can be skipped in the following sections.

Tokenization

Within the context of a query optimizer, *tokenization* is the process of breaking down the SQL statement into smaller units called *tokens*. Tokenization breaks down a query into keywords, identifiers, literals, operators, and punctuation symbols. The tokens are used to identify the structure and meaning of the query constituent parts in machine-usable format.

Tokenization is used by many RDBMS query optimizers but is not explicitly called out as a discrete component within Snowflake query optimization. It may be an inferred component within the Snowflake query optimization process and is mentioned to provide context for those migrating from legacy RDBMSs.

An alternative use of the word *tokenization* relates to both cybersecurity and substitution of data content with undecipherable tokens.

Parsing

Parsing is also referred to as *syntactic analysis* and is the action of analyzing the SQL statement structure or tokens. It validates the integrity of the query to ensure completeness and correctness prior to producing a parse tree from which intermediate code can be generated.

The parse tree provides the hierarchy and structure of the query along with all internal relationships and converts the form to Query Block Internal Representation (QBIR).

Semantic Analysis

Semantic analysis receives the QBIR and validates the structure matches both available and accessible Snowflake objects. We assume a lookup to FoundationDB is performed at this time. FoundationDB holds our Snowflake account metadata, that is, information about every object, relationship, and security feature. This is the catalog that documents and articulates your account.

Semantic analysis involves resolving object and attribute names, performs tag checking, expands referenced views, expands user-defined functions (UDFs), and expands common table expression (CTEs).

We anticipate additional checks are performed during semantic analysis including optional steps, which may be skipped.

We also understand this component performs entitlement checking to ensure only accessible objects are referenced and applies both row access policies and data masking policies. There may be additional functionality performed, but this list provides a flavor of known (or expected) capabilities delivered by this component.

Referential Integrity

An original design decision made by Snowflake was to allow referential integrity to be declared but not enforced. The only constraint enforced is NOT NULL. You can find further information at https://docs.snowflake.com/en/sql-reference/constraints-overview.

However, the forthcoming Unistore and hybrid tables change the Snowflake approach, at least for hybrid tables. It is not clear whether constraints previously not enforced will become optionally enforceable in the future for standard Snowflake tables. You can find further details on Unistore at https://www.snowflake.com/blog/introducing-unistore/ and https://www.snowflake.com/en/data-cloud/workloads/unistore/.

Regardless of the future state of constraints, I strongly recommend declaring constraints even if they are not enforced as their presence greatly assists data discovery via self-service tools, aids cataloging tooling, and is generally accepted as good practice. Some third-party tooling relies upon the presence of constraints to eliminate nonrequired tables prior to submitting queries to Snowflake.

Where possible, declaring constraints is good practice.

I also believe the presence of unenforced constraints informs query optimizer processing, but it is certain their presence is essential for hybrid tables, so you should adopt best practice wherever possible. You can find further information at https://docs.snowflake.com/en/sql-reference/constraints.

Logical Rewriter

After semantic analysis, the QBIR is passed to the Logical Rewriter where rules and algorithms are applied to restate the QBIR into an optimal internal representation. It is reasonable to assume the optimizer statistics (as listed in Chapter 1) inform the rules and algorithms. Furthermore, you can assume this step is a multipass process where many different QBIR representations are generated and compared to derive the optimal internal representation.

Micro-Partition Pruner

The optimal QBIR is received by the micro-partition pruner, which acts as the name suggests by invoking the pruner to exclude micro-partitions from consideration in resolving the eventual query result set. You can assume the optimizer statistics (as listed in Chapter 1) inform the micro-partition pruning strategy.

Micro-partition pruning occurs during several stages of the query plan generation and is implemented via the pruner, as explained in Chapter 3. For now, it is sufficient to understand micro-partition pruning conceptually within this overview.

Initial Plan Generation

After first-pass pruning has occurred, an initial execution plan is generated called the query plan (QP) internal representation, which is then passed to the plan rewriter.

We anticipate additional checks are performed during initial plan generation, including optional steps that may be skipped.

Plan Rewriter

Within the plan rewriter, a suite of rules applied to the QP causes rewrites, which may result in further micro-partition pruning implemented via the pruner shown separately.

You can assume additional checks are performed during plan rewrite including optional steps that may be skipped.

Cost-Based Join Ordering

Cost-based join ordering implies a rules-based approach to finalizing the QP. In truth, not much is known about this step; perhaps the clue is in the name, and the step simply orders the data access paths.

You can assume additional checks are performed during cost-based join ordering including optional steps that may be skipped.

Physical Query Plan

Finally, the physical query plan is delivered for execution. This is the optimal or "best" plan developed through application of all the prior steps.

The physical query plan is a directed acyclic graph (DAG) for which further information can be found here: `https://en.wikipedia.org/wiki/Directed_acyclic_graph`.

A DAG may be thought of as a flowchart with unidirectional links to branching logic where a determination is made to proceed or finish. There are no cycles or loops within the DAG, and it is not possible to follow a series of directed edges and return to the same node.

DAGs provide a useful way to represent dependencies, workflows, and hierarchical relationships between different elements in a system or problem.

Within the Snowflake physical query plan, the branching logic is one of the following:

- Operators that either process data or implement a feature such as aggregation, filtering, or summary.

- Links that are pipelines connecting operators or implementing parallelization features.

You can help the query optimizer by delivering the minimally simplest code. As it happens, the simplest code is most often the easiest to read and best laid out. Remember: smallest table first.

Query Execution

Data stored in a hybrid, columnar, compressed format lends itself to parallel processing. Snowflake processes queries using massively parallel processing (MPP) compute clusters where each node in the cluster stores a portion of the entire data set locally in columnar format.

In other words, if you can "chunk" data into discrete groups, each "chunk" can be processed independently by a separate processing unit.

Storing data in an organized manner improves clustering too; I will discuss this further in Chapter 4.

Warehouses

Each Snowflake node is a warehouse, and I discuss warehouses and their use later within this book. To not get bogged down in too much detail, for now just remember an extra small (XSmall) warehouse has eight CPUs and an associated cache, about 16 to 24 GB RAM, local SSD storage, and remote attached storage. For every T-shirt size we increase our warehouse to use, the number of CPUs doubles and memory increases too. These values will become very important as we progress through this book.

I use the term *execution unit* to respect the multithreaded/multicore/multimode operation of the underlying hardware; I don't always know the CSP hardware capabilities.

In the old world of on-prem databases, we had an appreciation of the physical hardware, including the number of available processing units, minus one for the operating system. Memory constraints also applied, and these limitations were judiciously managed by others. In the new world of cloud computing, we are abstracted away from the physical hardware and largely free to allocate resources on demand.

Without wanting to sound flippant, don't get absorbed in the details. Instead, you should accept there are some limitations to warehouses, but the elastic nature allows us to reconsider our approach by allocating resources on demand.

Single Instruction, Multiple Data (SIMD)

Snowflake query execution implements SIMD instructions, a technique optimized for data-level parallelism where each processing unit performs identical instructions on different data.

In Figure 2-3 we assume four processing units all performing the same "times two" operation against different data for two execution cycles.

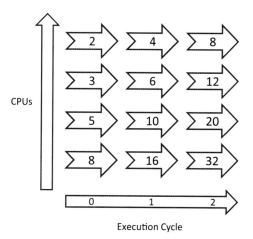

Figure 2-3. *SIMD example*

Compression

Snowflake query execution also makes use of compression; however, again available details are sparse. Compression may refer to in-memory compressed data, which is decompressed "on the fly" or to data stored in CPU caches, local SSD, and remote storage.

I know Snowflake applies a combination of compression methods including some tailored to the specific data type used. You can also be confident Snowflake provides the best compression available as part of their service and continually strives to improve their service.

Vectorization

The query execution engine is also vectorized, handling batches of a few thousand rows at a time. The actual batch size is unspecified, but given the propensity for programmers to prefer factors of 2, and speculation on my part, it may be in the range of 2,048 to 8,192 rows, and these rows will be in columnar format.

Query execution may also spill result sets to both local storage (SSD) and remote storage where result sets exceed CPU cache and allocated memory.

Flow Control

Two types of flow control model exist.

- **Pull-based:** The consumer continually polls for messages at the publisher.

- **Push-based:** The publisher pushes messages to the consumer as they become available.

The Snowflake query execution engine implements push-based flow control. As soon as results are available, they are pushed to consumers and further processed in a pipelined manner.

Note that after statement execution, summary information for it is available in `snowflake.account_usage.query_history` with a maximum latency of 45 minutes and visible for a year.

The performance profile provides more informative details and is retained for only two weeks, so analysis of long-running statements should be completed within this time to focus on optimization efforts, which assist with quicker execution and reduce Snowflake billing.

Summary

Snowflake implements a cost-based approach to delivering their query optimizer, which has many steps in common with other RDBMS vendors. Every RDBMS vendor implements bespoke optimizations, and Snowflake is no exception where these optimizations focus on satisfying edge cases and analytic specific features.

With a tight focus on delivering a robust optimizer that delivers stable execution plans, Snowflake deliberately removes levers and switches to influence system behavior and instead relies upon the built-in core query optimizer capability to handle as much as possible.

In this chapter I explained at a summary level the steps taken, from submitting a query through getting the corresponding result set. You then looked into the processing steps required to plan and optimize your query before execution.

You then looked at how a query is executed and began to see just how complex the query optimizer is. I also exposed some scenarios such as spills and parallelization where tuning will help.

Having established baseline information in preparation for looking deeper into performance tuning, I will cover query profiles in the next chapter.

CHAPTER 3

The Query Profiler

In Chapter 2, I showed how the query optimizer processes a SQL statement to produce a physical query plan. You will investigate how the query profiler operates by executing a query plan and generating execution statistics that expose various metrics.

This chapter initially focuses on the visual aspects of query profiling and later focuses on remediating issues.

The hands-on examples utilize TPC-H data supplied by Snowflake; you can find additional information about this data set at `https://docs.snowflake.com/en/user-guide/sample-data-tpch`. I assume you have access to a Snowflake account, but if not, a trial account is available at `www.snowflake.com`. Click the Start For Free button, and enter a few details to start your 30-day free trial.

Throughout this chapter, I will reference use cases to illustrate the query profiler behavior and later in this book will reference the same queries to demonstrate how you can identify and remediate performance issues.

Please note that some of the "bad" queries will consume all your credits; therefore, please read through this chapter carefully as I explain "why" and offer mitigating actions to prevent excessive credit consumption.

You may prefer to set `statement_timeout_in_seconds` in the current session to avoid overspend. In this example, you can set the timeout to 600 seconds (10 minutes).

```
ALTER SESSION SET statement_timeout_in_seconds = 600;
```

You can find further details at `https://docs.snowflake.com/en/sql-reference/parameters#statement-timeout-in-seconds`. You can find supplemental information at `https://community.snowflake.com/s/article/Parameter-STATEMENT-TIMEOUT-IN-SECONDS-covers-the-overall-time-of-query-execution`.

© Andrew Carruthers 2024
A. Carruthers, *Tuning the Snowflake Data Cloud*, https://doi.org/10.1007/979-8-8688-0379-6_3

Query Profile Overview

In this section, you will learn how to utilize SnowSight to access the query profiler. I am assuming your Snowflake account is available and ready for use.

In this section, you will use the imported share SNOWFLAKE_SAMPLE_DATA database, which is provisioned on account creation.

First, within your new worksheet, change the role to ACCOUNTADMIN:

```
USE ROLE accountadmin;
```

Your database tab should refresh to display the database SNOWFLAKE_SAMPLE_DATA.

Figure 3-1 shows the database SNOWFLAKE_SAMPLE_DATA within the database browser. Hover the mouse over the three dots [...] outlined in red in the figure; you'll see the pop-up window showing details of the database SNOWFLAKE_SAMPLE_DATA. You can see the imported database is a share.

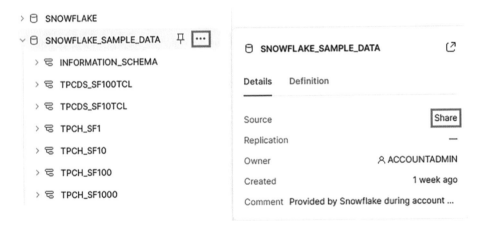

Figure 3-1. *SNOWFLAKE_SAMPLE_DATA database*

Many Snowflake accounts drop the imported SNOWFLAKE_SAMPLE_DATA database. If your account does not show SNOWFLAKE_SAMPLE_DATA within the database browser while using the role ACCOUNTADMIN, you may want to either create a new trial account or follow the instructions at https://docs.snowflake.com/en/user-guide/sample-data-using to reimport the dropped share.

Approach

You could simply run queries against `SNOWFLAKE_SAMPLE_DATA` using the `ACCOUNTADMIN` role. There are no special tuning optimizations for using the `ACCOUNTADMIN` role, but I prefer to develop my code as if it were to be deployed into production regardless of whether I throw the code away or retain it for later use.

I strongly encourage new developers to treat all code as if it were going to be reviewed and tested for production, and part of my approach is to insist on properly formatted and commented code. I prefer keywords to shout at me and for all other words to be in lowercase. Some people agree, others disagree, and everyone is entitled to their view.

Regardless, you will reuse some code as you progress through this book, so let's assume you are aiming for a production release. More will be revealed later as you progress through the chapters.

Setup

Using the new worksheet, let's create an environment in which to develop the code and build components for later reuse. You will start with creating a database, several warehouses, and a role, and then you will enable access to the Account Usage Store.

You will reuse the environment you are about to create throughout this book.

Snowflake makes reference to the Account Usage Store, which is the imported share visible within the database share referred to as SNOWFLAKE. Figure 3-1 shows this imported database above SNOWFLAKE_SAMPLE_DATA.

As SNOWFLAKE_SAMPLE_DATA is an imported share, you cannot modify the contents. To investigate the query profiler, let's create an environment for which you also supply an accompanying script.

You will first declare identifiers to be used throughout this chapter. Note that you may need to rerun these declarations when you open the browser session again.

```
SET tpc_owner_role   = 'tpc_owner_role';
SET tpc_warehouse_XS = 'tpc_wh_xsmall';
SET tpc_warehouse_S  = 'tpc_wh_small';
```

```
SET tpc_warehouse_M  = 'tpc_wh_medium';
SET tpc_warehouse_L  = 'tpc_wh_large';
SET tpc_warehouse_XL = 'tpc_wh_xlarge';
SET tpc_database     = 'tpc';
SET tpc_owner_schema = 'tpc.tpc_owner';
```

You can use the <u>sysadmin</u> role to create first-order database objects such as databases, schemas, warehouses, and shares.

Create a database called <u>TPC</u> and a schema within <u>TPC</u> called tpc_owner.

```
USE ROLE sysadmin;

CREATE OR REPLACE DATABASE IDENTIFIER ( $tpc_database ) DATA_RETENTION_
TIME_IN_DAYS = 90;

CREATE OR REPLACE SCHEMA IDENTIFIER ( $tpc_owner_schema   );
```

Now create five warehouses of increasing size up to XL. You could add larger warehouses from 2XL up to 6XL, but these five declared warehouses are sufficient for our purposes right now.

Remember, unless explicitly declared with "INITIALLY_SUSPENDED" = TRUE, warehouses run when they are declared and run when invoked to process a query.

For the warehouse declarations, set the default clustering to 1. I will explain warehouse clustering later as the subject is worthy of its own chapter.

```
CREATE OR REPLACE WAREHOUSE IDENTIFIER ( $tpc_warehouse_XS ) WITH
WAREHOUSE_SIZE       = 'X-SMALL'
AUTO_SUSPEND         = 60
AUTO_RESUME          = TRUE
MIN_CLUSTER_COUNT    = 1
MAX_CLUSTER_COUNT    = 1
SCALING_POLICY       = 'STANDARD'
INITIALLY_SUSPENDED = TRUE;

CREATE OR REPLACE WAREHOUSE IDENTIFIER ( $tpc_warehouse_S ) WITH
WAREHOUSE_SIZE       = 'SMALL'
AUTO_SUSPEND         = 60
AUTO_RESUME          = TRUE
MIN_CLUSTER_COUNT    = 1
```

```
MAX_CLUSTER_COUNT    = 1
SCALING_POLICY       = 'STANDARD'
INITIALLY_SUSPENDED = TRUE;

CREATE OR REPLACE WAREHOUSE IDENTIFIER ( $tpc_warehouse_M ) WITH
WAREHOUSE_SIZE       = 'MEDIUM'
AUTO_SUSPEND         = 60
AUTO_RESUME          = TRUE
MIN_CLUSTER_COUNT    = 1
MAX_CLUSTER_COUNT    = 1
SCALING_POLICY       = 'STANDARD'
INITIALLY_SUSPENDED = TRUE;

CREATE OR REPLACE WAREHOUSE IDENTIFIER ( $tpc_warehouse_L ) WITH
WAREHOUSE_SIZE       = 'LARGE'
AUTO_SUSPEND         = 60
AUTO_RESUME          = TRUE
MIN_CLUSTER_COUNT    = 1
MAX_CLUSTER_COUNT    = 1
SCALING_POLICY       = 'STANDARD'
INITIALLY_SUSPENDED = TRUE;

CREATE OR REPLACE WAREHOUSE IDENTIFIER ( $tpc_warehouse_XL ) WITH
WAREHOUSE_SIZE       = 'X-LARGE'
AUTO_SUSPEND         = 60
AUTO_RESUME          = TRUE
MIN_CLUSTER_COUNT    = 1
MAX_CLUSTER_COUNT    = 1
SCALING_POLICY       = 'STANDARD'
INITIALLY_SUSPENDED = TRUE;
```

You can use the <u>securityadmin</u> role to create roles and add object entitlements to roles. You can first create a new role called tpc_owner_role.

```
USE ROLE securityadmin;

CREATE OR REPLACE ROLE IDENTIFIER ( $tpc_owner_role );
```

You may prefer to vary the entitlements granted to roles; this is a simple template for you to later expand.

Then grant entitlements to the role called `tpc_owner_role` starting with database entitlements.

```
GRANT IMPORTED PRIVILEGES ON DATABASE snowflake TO ROLE IDENTIFIER
( $tpc_owner_role );

GRANT USAGE   ON DATABASE  IDENTIFIER ( $tpc_database      ) TO ROLE
IDENTIFIER ( $tpc_owner_role  );
```

Add warehouse entitlements.

```
GRANT USAGE   ON WAREHOUSE IDENTIFIER ( $tpc_warehouse_XS  ) TO ROLE
IDENTIFIER ( $tpc_owner_role  );
GRANT OPERATE ON WAREHOUSE IDENTIFIER ( $tpc_warehouse_XS  ) TO ROLE
IDENTIFIER ( $tpc_owner_role  );
GRANT USAGE   ON WAREHOUSE IDENTIFIER ( $tpc_warehouse_S   ) TO ROLE
IDENTIFIER ( $tpc_owner_role  );
GRANT OPERATE ON WAREHOUSE IDENTIFIER ( $tpc_warehouse_S   ) TO ROLE
IDENTIFIER ( $tpc_owner_role  );
GRANT USAGE   ON WAREHOUSE IDENTIFIER ( $tpc_warehouse_M   ) TO ROLE
IDENTIFIER ( $tpc_owner_role  );
GRANT OPERATE ON WAREHOUSE IDENTIFIER ( $tpc_warehouse_M   ) TO ROLE
IDENTIFIER ( $tpc_owner_role  );
GRANT USAGE   ON WAREHOUSE IDENTIFIER ( $tpc_warehouse_L   ) TO ROLE
IDENTIFIER ( $tpc_owner_role  );
GRANT OPERATE ON WAREHOUSE IDENTIFIER ( $tpc_warehouse_L   ) TO ROLE
IDENTIFIER ( $tpc_owner_role  );
GRANT USAGE   ON WAREHOUSE IDENTIFIER ( $tpc_warehouse_XL  ) TO ROLE
IDENTIFIER ( $tpc_owner_role  );
GRANT OPERATE ON WAREHOUSE IDENTIFIER ( $tpc_warehouse_XL  ) TO ROLE
IDENTIFIER ( $tpc_owner_role  );
```

Add schema entitlements.

```
GRANT USAGE   ON SCHEMA    IDENTIFIER ( $tpc_owner_schema   ) TO ROLE
IDENTIFIER ( $tpc_owner_role  );
```

Add object entitlements, and note the inclusion of dynamic tables, which are currently in public preview.

```
GRANT USAGE                          ON SCHEMA IDENTIFIER ( $tpc_owner_
schema    ) TO ROLE IDENTIFIER ( $tpc_owner_role );
GRANT MONITOR                        ON SCHEMA IDENTIFIER ( $tpc_owner_
schema    ) TO ROLE IDENTIFIER ( $tpc_owner_role );
GRANT MODIFY                         ON SCHEMA IDENTIFIER ( $tpc_owner_
schema    ) TO ROLE IDENTIFIER ( $tpc_owner_role );
GRANT CREATE TABLE                   ON SCHEMA IDENTIFIER ( $tpc_owner_
schema    ) TO ROLE IDENTIFIER ( $tpc_owner_role );
GRANT CREATE DYNAMIC TABLE      ON SCHEMA IDENTIFIER ( $tpc_owner_
schema    ) TO ROLE IDENTIFIER ( $tpc_owner_role );
GRANT CREATE VIEW                    ON SCHEMA IDENTIFIER ( $tpc_owner_
schema    ) TO ROLE IDENTIFIER ( $tpc_owner_role );
GRANT CREATE SEQUENCE                ON SCHEMA IDENTIFIER ( $tpc_owner_
schema    ) TO ROLE IDENTIFIER ( $tpc_owner_role );
GRANT CREATE FUNCTION                ON SCHEMA IDENTIFIER ( $tpc_owner_
schema    ) TO ROLE IDENTIFIER ( $tpc_owner_role );
GRANT CREATE PROCEDURE               ON SCHEMA IDENTIFIER ( $tpc_owner_
schema    ) TO ROLE IDENTIFIER ( $tpc_owner_role );
GRANT CREATE STREAM                  ON SCHEMA IDENTIFIER ( $tpc_owner_
schema    ) TO ROLE IDENTIFIER ( $tpc_owner_role );
GRANT CREATE MATERIALIZED VIEW  ON SCHEMA IDENTIFIER ( $tpc_owner_
schema    ) TO ROLE IDENTIFIER ( $tpc_owner_role );
GRANT CREATE FILE FORMAT             ON SCHEMA IDENTIFIER ( $tpc_owner_
schema    ) TO ROLE IDENTIFIER ( $tpc_owner_role );
```

Before you can use the new tpc_owner_role, you must grant the role to yourself.

```
GRANT ROLE IDENTIFIER ( $tpc_owner_role ) TO USER <Your Name Here>;
```

TPC Data Model

Figure 3-2 represents the TPC data model taken from the Snowflake sample TPCH data found at https://docs.snowflake.com/en/user-guide/sample-data-tpch.

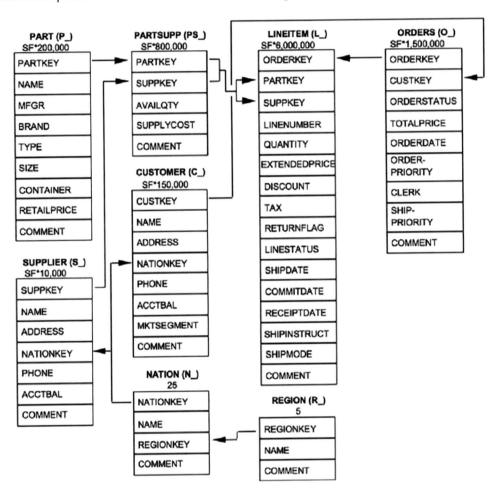

Figure 3-2. *TPC-H entity relationship diagram*

You will next copy tables after which you will begin your investigation of query profile behavior.

Initial Population

Within this section you will copy over <u>SNOWFLAKE_SAMPLE_DATA</u> from the TPCH_ SF1000 schema, which contains the largest sample datasets.

Before you do anything, you must change to `tpc_owner_role`.

```
USE ROLE      IDENTIFIER ( $tpc_owner_role   );
USE DATABASE  IDENTIFIER ( $tpc_database     );
USE SCHEMA    IDENTIFIER ( $tpc_owner_schema );
USE WAREHOUSE IDENTIFIER ( $tpc_warehouse_xs );
```

I (almost) always explicitly set my execution context by setting the database, schema, warehouse, and role. I strongly recommend this approach as a best practice. I have lost count of the number of times I ended up using the wrong role, wasting time and effort, so I make a practice of explicitly setting up each environment at the outset.

When asked to investigate queries, I insist upon having the context along with the query.

The current execution context can be derived by this query:

```
SELECT current_role(),
       current_warehouse(),
       current_database(),
       current_schema();
```

As you know from Chapter 2, even though constraints are not enforced, the query optimizer can use declared constraints.

To ensure you have equivalence between your data source and copied data, you must check whether constraints have been declared on the source tables. You perform this check by investigating `information_schema` for declared constraints. In this example, you are looking for referential integrity constraints declared within any schema in the SNOWFLAKE_SAMPLE_DATA database:

```
SELECT DISTINCT unique_constraint_schema
FROM    snowflake_sample_data.information_schema.referential_constraints;
```

You should expect to see a single row TPCDS_SF100TCL indicating the chosen schema TPCH_SF1000 does not have any constraints declared.

Having previously set the environment, let's copy the tables across. Wherever possible, I use self-generating SQL. Note the addition of the _baseline suffix as you will be creating more objects later.

```
SELECT 'CREATE OR REPLACE TABLE '||
       LOWER ( table_name )||'_baseline'||
       ' AS SELECT * FROM snowflake_sample_data.tpch_sf1000.'||
       LOWER ( table_name )||';'
FROM   snowflake_sample_data.information_schema.tables
WHERE  table_schema = 'TPCH_SF1000';
```

Remember that we are using an X-Small warehouse; therefore, runtimes will be considerable. Cut and paste the generated output back into SnowSight.

Before executing the generated code and because this is a book on performance tuning, let's look at runtimes for different size warehouses. To save you the expense of running incorrectly sized warehouses, I have run Create Table As SELECT (CTAS) benchmarks using code generated from the previous query. The timings shown in Table 3-1 are in seconds. Note there may be some small variances in your runtimes should you choose to repeat the tests.

Table 3-1. *TPC Baseline Table Copy Times and Data Volumes*

Table/Warehouse	X-Small	Small	Medium	Large	X-Large	Row Count
customer_baseline	80	44	25	15	11	150000000
lineitem_baseline	2195	1099	561	296	165	5999989709
nation_baseline	1	1	1	1	1	25
partsupp_baseline	291	151	79	44	25	800000000
region_baseline	1	1	1	1	1	5
orders_baseline	543	279	147	80	48	1500000000
part_baseline	81	42	24	16	11	200000000
supplier_baseline	9.1	9	9	4	9	10000000

We recommend using an XL warehouse for high-volume data copies, as indicated by the timings shown in Table 3-1.

The most important lessons from Table 3-1 are to know your data volumes and to size your warehouse accordingly. I will discuss costs later.

I will return to warehouse tuning later in this book as there is much more to unpack and the subject deserves a chapter to itself.

You must clear out the cache before rerunning or your performance figures will be incorrect. To do this, you set your session to ignore cached results causing every SQL statement to be executed.

```
ALTER SESSION SET use_cached_result = FALSE;
```

You can find further information on disabling cached results at https://docs. snowflake.com/en/sql-reference/parameters#use-cached-result.

You can find supplemental information at https://docs.snowflake.com/en/user-guide/querying-persisted-results.

Disabling cached results is for performance tuning only.

*** Do not disable cached results in your production code. ***

Then declare your chosen warehouse.

```
USE WAREHOUSE IDENTIFIER ( $tpc_warehouse_xs );
```

Warehouse declaration does not clear out the warehouse cache, so you must suspend and restart your warehouse, which also aborts all active SQL statements.

```
ALTER WAREHOUSE IDENTIFIER ( $tpc_warehouse_xs ) SUSPEND;
```

```
ALTER WAREHOUSE IDENTIFIER ( $tpc_warehouse_xs ) RESUME;
```

*** Warehouse suspension is for performance tuning only. ***

Now execute your baseline table creation statements and set the warehouse once for each statement group according to the optimal size shown in Table 3-1.

```
USE WAREHOUSE IDENTIFIER ( $tpc_warehouse_xl );

CREATE OR REPLACE TABLE customer_baseline
AS SELECT * FROM snowflake_sample_data.tpch_sf1000.customer;
CREATE OR REPLACE TABLE lineitem_baseline
AS SELECT * FROM snowflake_sample_data.tpch_sf1000.lineitem;
CREATE OR REPLACE TABLE partsupp_baseline
AS SELECT * FROM snowflake_sample_data.tpch_sf1000.partsupp;
CREATE OR REPLACE TABLE orders_baseline
AS SELECT * FROM snowflake_sample_data.tpch_sf1000.orders;
CREATE OR REPLACE TABLE part_baseline
AS SELECT * FROM snowflake_sample_data.tpch_sf1000.part;

USE WAREHOUSE IDENTIFIER ( $tpc_warehouse_xs );

CREATE OR REPLACE TABLE nation_baseline
AS SELECT * FROM snowflake_sample_data.tpch_sf1000.nation;
CREATE OR REPLACE TABLE region_baseline
AS SELECT * FROM snowflake_sample_data.tpch_sf1000.region;
CREATE OR REPLACE TABLE supplier_baseline
AS SELECT * FROM snowflake_sample_data.tpch_sf1000.supplier;
```

Having briefly demonstrated how warehouse sizing affects performance while creating a test dataset, I will move on to show query profiles.

Query Profiles

In this section, I discuss how to access query profiles using the options available in SnowSight. I then show how to develop a simple query using the TPC baseline data to provide a more detailed explanation of how the query profiler behaves.

I stressed the importance of sizing the warehouse appropriately, but you may find in practice the declared warehouse is not used. This may happen for several reasons.

- The requested results are satisfied from the query cache.

- The query may be satisfied from the metadata repository.

You can now begin investigating query profile characteristics.

Accessing Query Profiles

Every query has a profile; you can access query profiles in several ways depending upon your starting point.

I next explain where and how to access query profile information. Note that I am introducing the topic to inform you how to access information and not jumping into the specifics of the subject yet.

Running the Query

Figure 3-3 shows a partial screenshot for a currently executing query. To see the query profile discussed next, click the text next to the "ID" label, which opens a new tab in the browser.

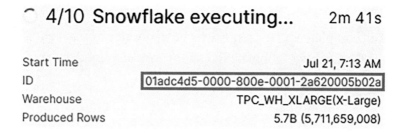

Figure 3-3. *Query profile from running a query*

To execute a query, you will see the available information only.

Completed Query

For a query that has completed execution, you can access the query profile by clicking the text next to the "Query ID" label, as shown in Figure 3-4, which opens a new tab in the browser.

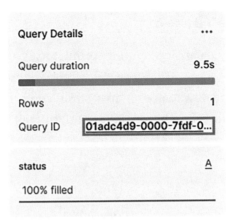

Figure 3-4. *Query profile from the completed query*

Query History

You can access all complete and currently executing queries by navigating to the Activity ➤ Query History page where all the queries are displayed ordered by execution timestamp in descending order, as shown in Figure 3-5.

Figure 3-5. *Accessing the query history*

From the query history page, select the desired query, and then select the Query Profile tab.

Get Query Operator Stats

Snowflake has released a new system function called GET_QUERY_OPERATOR_ STATS() that is now in Generally Available (GA) status, which I will discuss later. You can find further information at https://docs.snowflake.com/en/sql-reference/functions/get_query_operator_stats.

Having identified various access paths to viewing query profiles, let's create an example query to work through the query profiler.

Example Query

Using our newly established baseline data set, let's create and execute an example query and then investigate the query profile. From the TPC ERD provided in Figure 3-2 and substituting the baseline tables, you will use the PARTSUPP_BASELINE, PART_BASELINE, and SUPPLIER_BASELINE tables initially.

As before, you can establish the good practice of clearing the cache before testing.

```
ALTER SESSION SET use_cached_result = FALSE;
```

Expected Result Count

In line with my earlier recommendation, you should know your expected data volumes, so let's do this first, and from Table 3-1 you have identified the table row counts. Therefore, set your warehouse accordingly.

```
USE WAREHOUSE IDENTIFIER ( $tpc_warehouse_xl );
```

Now execute your query.

```
SELECT count(1)
FROM   partsupp_baseline ps,
       part_baseline     p,
       supplier_baseline s
WHERE  ps.ps_partkey      = p.p_partkey
AND    ps.ps_suppkey      = s.s_suppkey;
```

Your query should return a record count of 800,000,000 in less than six seconds. If you had used an X-Small warehouse, your query runtime would have been around 43 seconds.

Failure to set warehouse size correctly will result in excessive consumption charges.

Cross-referencing the record count to the query execution times in Table 3-1, you can see that a reasonable starting point for the warehouse is Large. You can adjust this up or down according to actual performance, an exercise I will leave for you.

Don't be frightened of using Large or bigger warehouses, but they can be more time and cost effective than using a smaller warehouse.

```
USE WAREHOUSE IDENTIFIER ( $tpc_warehouse_1 );
```

Suspend and restart your warehouse.

```
ALTER WAREHOUSE IDENTIFIER ( $tpc_warehouse_1 ) SUSPEND;

ALTER WAREHOUSE IDENTIFIER ( $tpc_warehouse_1 ) RESUME;
```

Developing an Example Query

Having set your warehouse and assuming your context (database, schema, and role) has not changed, let's define an imaginary scenario using the TPC data to base your example query on.

You might imagine yourself working within the IT department of a machine parts supplier. You hold stock purchased over time; therefore, you will have multiple records for a single named part. Because of currency fluctuations, the purchase price has varied. Your objective is to identify the costliest stock and sell this first for accounting reasons.

Let's first create a view called v_supplier_part to both denormalize the data model and simplify your data access path. You can remove extraneous attributes to keep the query small.

```
CREATE OR REPLACE VIEW v_supplier_part COPY GRANTS
AS
SELECT p.p_name            AS part_name,
       p.p_retailprice     AS retail_price,
       ps.ps_supplycost    AS supply_cost,
       ps.ps_availqty      AS available_quantity,
       ps.ps_supplycost / ps.ps_availqty    AS unit_price,
       s.s_acctbal         AS supplier_account_balance,
       s.s_name            AS supplier_name
FROM   partsupp_baseline ps,
       part_baseline      p,
       supplier_baseline s
WHERE  ps.ps_partkey     = p.p_partkey
AND    ps.ps_suppkey     = s.s_suppkey;
```

Then run a simple query to access v_supplier_part to produce a query profile.

```
SELECT part_name,
       available_quantity      AS avail_qty,
       unit_price,
       supplier_account_balance AS acct_bal,
       supplier_name
FROM   v_supplier_part
ORDER BY part_name         ASC,
         unit_price        DESC
LIMIT 10;
```

Let's examine the query profile; I showed you how to access the query profile earlier, but as a quick reminder, Figure 3-6 shows the query and result set in context along with highlighting the query ID, which should be clicked.

```
SELECT part_name,
       available_quantity       AS avail_qty,
       unit_price,
       supplier_account_balance AS acct_bal,
       supplier_name
FROM   v_supplier_part
ORDER BY part_name        ASC,
         unit_price       DESC
LIMIT 10;
```

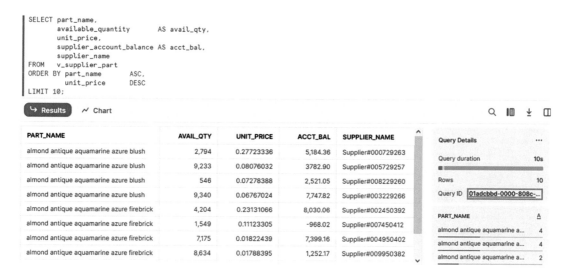

Figure 3-6. *Sample query results*

You should also note that the query duration is 10 seconds.

Profiling Your Example Query

After clicking the query ID, a new tab will open in your browser where for the first time you will see the query profile, as shown in Figure 3-7. Note that the color coding (for the PDF version) is mine.

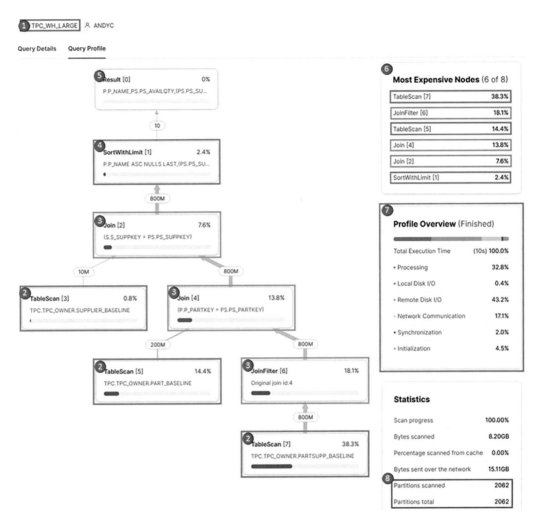

Figure 3-7. *Sample query profile*

The first point to note is that view expansion has occurred. The example query references the view v_supplier_part, which the query optimizer has expanded into its constituent tables and join criteria.

As you can see in Figure 3-7, there is a lot happening. Let's break the profile down into its constituent parts. Here's an explanation of each step:

1. The warehouse used to execute the sample query is displayed. I prefer to put the warehouse size into my naming convention to provide context, but as I will discuss later, there is no guarantee the name correlates to the actual warehouse size.

2. The leaf nodes of the tree are the physical objects, in this example, tables, which are expensive to access (see step 6).

3. Joins and join filters resolve both the total volume of data selected and the attributes returned. In this example, there are no filters to subset the results. The example query joined two source tables (part_baseline and supplier_baseline) with an intersection table (partsupp_baseline).

4. The ORDER BY and LIMIT clauses are evaluated last and shown as SortWithLimit. ORDER BY always results in a Sort operation.

5. The result set limited to 10 records is returned to SnowSight; note this is a parallel operation.

6. Here you can see the ordered list of operations from the most expensive to the least expensive with table access and joins being the most expensive.

7. The profile overview shows where the most effort is expended when executing the query. You should pay particular attention to local disk I/O and remote disk I/O when performance tuning.

8. Where "Partitions scanned" is less than "Partitions total," you know partition pruning has occurred. In the example query without filters, you should not be surprised to see these values are identical.

One further point to note: If you look carefully at the query profile, between the nodes you will see the number of rows input to each parent node, with proportionally thicker lines indicating record volume.

Overlaying the query parsing order from Chapter 1 allows you to visualize the query profile in context. Figure 3-8 overlays the query parsing order onto the query profile.

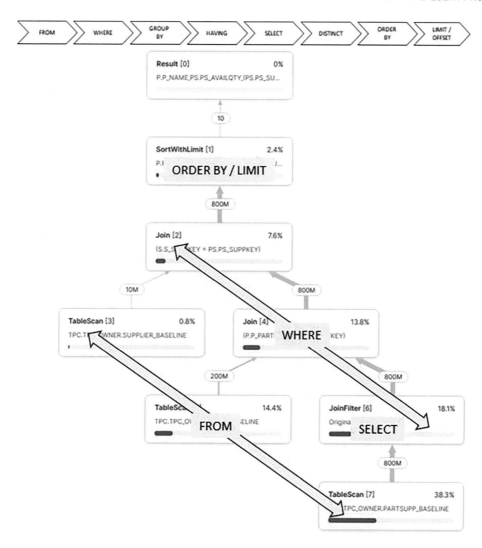

Figure 3-8.   *Sample query with parsing order*

You will now briefly investigate the creation and use of a *dynamic table* (previously called *materialized table*), a new feature currently in public preview.

Materializing Your Example Query

You can replace the earlier example query with a dynamic table. The purpose of this section is to offer an alternative performance tuning approach.

Within Snowflake, a materialized view can be declared only on a single table and is essentially a way to declare alternative cluster keys on a base table. Using materialized views facilitates micro-partition pruning via aggregation, a topic I will discuss in Chapter 4.

In Snowflake, materialized views differ significantly from legacy RDBMS equivalent implementations. Be prepared to set aside any assumptions on materialized view implementation and capability. Within Snowflake, materialized views incur maintenance, runtime, and storage costs. Before implementing and using materialized views, you must strike a balance to ensure optimally cost-effective solutions are developed and delivered. Similar considerations apply to the use of dynamic tables, discussed next.

You can find further information on materialized views at `https://docs.snowflake.com/en/user-guide/views-materialized`.

For those familiar with legacy RDBMSs, comparable functionality to that provided by dynamic tables has been around for a long time, where the ability to join multiple tables and create a table-like object offers significant performance improvements. A dynamic table maintains the result set from a query on a scheduled timer. Think of a dynamic table as the conflation of a stream, a task, and, in this example, a multitable-based view.

There is a trade-off: you will incur additional storage costs along with serverless compute costs for provisioning dynamic tables.

In this example, you will use an X-Small warehouse for the periodic refreshes. This was chosen as a reasonable compromise because you will consume one credit every time your X-Small warehouse runs for an hour. You may want to investigate using a different size warehouse.

```
CREATE OR REPLACE DYNAMIC TABLE dt_supplier_part COPY GRANTS
TARGET_LAG = '30 MINUTES'
WAREHOUSE  = tpc_wh_xsmall
AS
SELECT p.p_name          AS part_name,
       p.p_retailprice   AS retail_price,
       ps.ps_supplycost  AS supply_cost,
       ps.ps_availqty    AS available_quantity,
```

```
        ps.ps_supplycost / ps.ps_availqty      AS unit_price,
        s.s_acctbal          AS supplier_account_balance,
        s.s_name             AS supplier_name
FROM    partsupp_baseline ps,
        part_baseline     p,
        supplier_baseline s
WHERE   ps.ps_partkey      = p.p_partkey
AND     ps.ps_suppkey      = s.s_suppkey;
```

When the dynamic table is declared, the <u>TARGET LAG</u> value dictates the first runtime; in this example, 30 minutes will elapse before the first refresh.

You must now resume the dynamic table.

```
ALTER DYNAMIC TABLE dt_supplier_part RESUME;
```

If you see the error message "Dynamic Table 'TPC.TPC_OWNER.DT_SUPPLIER_PART' is not initialized. Please run a manual refresh or wait for a scheduled refresh before querying." then you must refresh the dynamic table to force a refresh.

```
ALTER DYNAMIC TABLE dt_supplier_part REFRESH;
```

On completion you should see status information similar to that shown in Figure 3-9.

dt_name	...	statistics	refreshed_dt_count	data_timestamp
TPC.TPC_OWNER.DT_SUPPLIER_PART		{"insertedRows":800000000,"copiedRows":0,"deletedRows":0}	1	1,690,405,015,284

Figure 3-9. *Dynamic table refresh output*

After refreshing the dynamic table, you can clear the cache.

```
ALTER SESSION SET use_cached_result = FALSE;
```

Then ensure you are using a Large warehouse to guarantee comparable execution context as before.

```
USE WAREHOUSE IDENTIFIER ( $tpc_warehouse_l );
```

Suspend and restart the warehouse.

```
ALTER WAREHOUSE IDENTIFIER ( $tpc_warehouse_l ) SUSPEND;
```

```
ALTER WAREHOUSE IDENTIFIER ( $tpc_warehouse_l ) RESUME;
```

Then run the same simple query accessing dt_supplier_part as shown earlier to generate a query profile.

```
SELECT part_name,
       available_quantity      AS avail_qty,
       unit_price,
       supplier_account_balance AS acct_bal,
       supplier_name
FROM   dt_supplier_part
ORDER BY part_name         ASC,
         unit_price        DESC
LIMIT 10;
```

Figure 3-10 shows the reduction in query duration from 10 seconds down to 4 seconds.

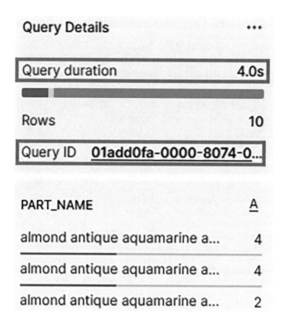

Figure 3-10. *Sample query refactored for dynamic table*

As you should expect, the query profile now references a single object, the new dynamic table dt_supplier_part, as shown in Figure 3-11.

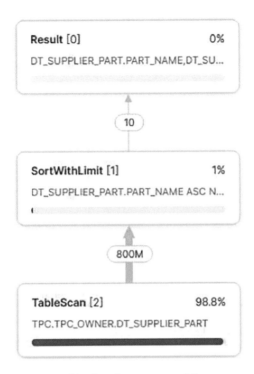

Figure 3-11. *Sample query profile for dynamic table*

Now that you have an understanding of what a query profile looks like and the relationship to how the query is parsed, let's investigate how to optimize the query profiles.

A Good Query Profile

In this section, you must first understand what a "good" query profile looks like and explain why a query profile is considered "good." You then demonstrate what a "bad" query profile looks like and demonstrate how to identify what constitutes "bad."

Snowflake optimizer join order heuristics are optimized for star schemas.

You begin by referencing the now familiar first query profile as this is a "good" example.

Knowing the Snowflake optimizer join order heuristics are optimized for star schemas allows you to differentiate "good" from "bad" query profiles. Using the familiar query profile, overlaid with new terminology "Build" and "Probe," as shown in Figure 3-12, is further explained next.

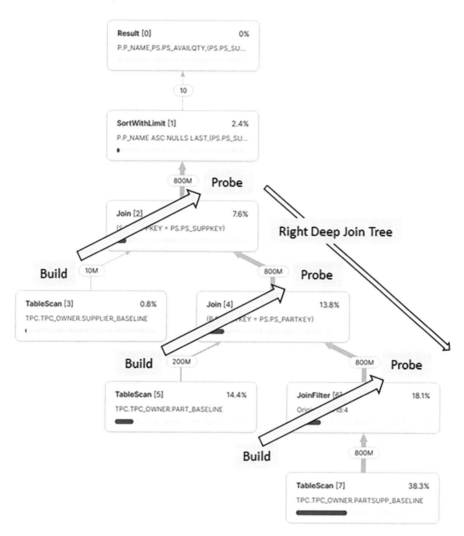

Figure 3-12. *Build, probe, and right deep tree*

Build Side

Build-side operations complete first, where small tables or dimensions are preferred.

Snowflake creates hash tables on the probe side in preparation for joining data sets. As illustrated in Figure 3-12, the order in which you join tables has significance. The optimizer prefers the table returning the largest data set (partsupp_baseline 800M records) at the bottom right and the table returning the smallest data set at the top left (supplier_baseline 10M records). I will further discuss join ordering later when considering performance tuning.

When the build side is larger than the probe side, performance is usually slower.

Probe Side

Probe-side operations are where large tables, result sets, or facts are preferred. Using small build-side data sets facilitates parallelization and optimal use of cluster memory where our hybrid columnar compressed data is held. As you can see in Figure 3-12, all build and probe operations are optimally positioned within the query profile.

Right Deep Join Tree

Seventy percent of people prefer visual representations, whereas 30 percent prefer textual representations of information. As advertising executive Frederick R. Barnard stated, "One picture is worth a thousand words." I agree and further propose the visual representation of a query profile readily allows the viewer to determine patterns that indicate "good" or "bad."

Figure 3-12 shows a right deep join tree, which consumes less warehouse memory and increases parallel processing options.

In my estimation, I consider the query profile tree and profile overview provide invaluable tools for performance tuning queries and strongly recommend a firm grasp of these fundamentals will serve you well.

Bloom Filter

Snowflake implements Bloom filters for probabilistic (not deterministic) testing of whether an element exists within a set of data and may return one of two outcomes.

- The element is possibly in the set of data and may contain false-positives.

- The element is definitely not in the set of data and enables pruning, leading to improved query performance.

A full explanation of Bloom filters is beyond the scope of this book; you can find more information at `https://en.wikipedia.org/wiki/Bloom_filter`.

Explain Plan

Before you learn about "bad" query profiles, you will investigate tooling provided to return—but not invoke—the execution plan for the current statement. To do this, Snowflake provides the `EXPLAIN` keyword, which may be prefixed to any `SELECT` statement, and by so doing, the query is only evaluated. With the query plan available, you can examine the quality of the execution plan.

It is important to note `EXPLAIN` is a metadata operation and therefore does not require a warehouse. However, like all other metadata operations, cloud service credits are consumed.

You will use `EXPLAIN` as you progress through the remainder of this chapter to investigate and identify both poorly constructed and badly executing queries.

To illustrate how `EXPLAIN` works, let's reuse the earlier known-good example query referencing `v_supplier_part`. In this example, you will request the `TABULAR` output but might instead prefer JSON output.

```
EXPLAIN USING TABULAR
SELECT part_name,
       available_quantity      AS avail_qty,
       unit_price,
       supplier_account_balance AS acct_bal,
       supplier_name
```

```
FROM    v_supplier_part
ORDER BY part_name        ASC,
         unit_price       DESC
LIMIT 10;
```

Figure 3-13 shows sample explain plan output noting further information is available scrolling off the right of the screen (not shown).

step	id	parent	operation	objects	alias	expressions
null	null	null	GlobalStats	null	null	null
1	0	null	Result	null	null	P.P_NAME, PS.PS_AVAILQTY, (PS.PS_SL
1	1	0	SortWithLimit	null	null	sortKey: [P.P_NAME ASC NULLS LAST,
1	2	1	InnerJoin	null	null	joinKey: (S.S_SUPPKEY = PS.PS_SUPPK
1	3	2	TableScan	TPC.TPC_OWNER.SUPPLIER_BASELINE	S	S_SUPPKEY, S_NAME, S_ACCTBAL
1	4	2	InnerJoin	null	null	joinKey: (P.P_PARTKEY = PS.PS_PARTKE
1	5	4	TableScan	TPC.TPC_OWNER.PART_BASELINE	P	P_PARTKEY, P_NAME
1	6	4	JoinFilter	null	null	joinKey: (S.S_SUPPKEY = PS.PS_SUPPK
1	7	6	TableScan	TPC.TPC_OWNER.PARTSUPP_BASELINE	PS	PS_PARTKEY, PS_SUPPKEY, PS_AVAILQ

Figure 3-13. *Explain plan output*

Interestingly, Snowflake also provides a query ID and link next to the explain plan output; however, no query profile is available. The absence of a query profile is due to the query not having been executed.

Snowflake also provides functions to convert EXPLAIN JSON to text. You can find further details at `https://docs.snowflake.com/en/sql-reference/functions/system_explain_plan_json`.

EXPLAIN is very useful when developing new queries. Having the capability for Snowflake to generate a query profile before execution can save time and prevent expensive mistakes. I suggest all unit tests include EXPLAIN output, and you could go further by profiling every SQL statement as part of the continuous integration testing and scan the output for keywords.

You can find further details for EXPLAIN at `https://docs.snowflake.com/en/sql-reference/sql/explain`.

GET_QUERY_OPERATOR_STATS

GET_QUERY_OPERATOR_STATS returns query operator information for completed queries. You will use this new table function to later programmatically identify rogue queries.

GET_QUERY_OPERATOR_STATS is limited to queries executed in the past 14 days.

For immediate results, you might prefer to use last_query_id() to identify information from the most recently run SQL statement. Note that GET_QUERY_OPERATOR_ STATS may return OPERATOR_TYPE of QUERY RESULT REUSE, which indicates the source query profile is inaccessible. In this example the query optimizer determined to use the result cache.

The following is the general form of this query:

```
SELECT <attributes>
FROM    TABLE ( get_query_operator_stats(<your value here>))
WHERE   <predicates>
ORDER BY <ordering>;
```

GET_QUERY_OPERATOR_STATS accepts a single value, which must be one of the following:

- The value returned by last_query_id()

- A session variable containing a valid query_id

- A string literal set to valid query_id

- The query_id values used throughout this book will vary when executed against your Snowflake account.

You will return to GET_QUERY_OPERATOR_STATS throughout the rest of this book. You can find further information at https://docs.snowflake.com/en/sql-reference/ functions/get_query_operator_stats.

Bad Query Profiles

Unfortunately, I see far more "bad" query profiles than "good" query profiles. One good reason for writing this book is to impart sufficient information to developers to enable them to identify "bad" query profiles. The very best developers check each query profile for optimal behavior and performance by both unit testing and checking query profiles before submitting their code for promotion into production systems.

However, performance tuning is often an after-thought. Our hope is the information found within this book will greatly assist and reduce the time it takes to identify and remediate performance issues.

Please do not execute the queries within this section. Most will consume significant credits along with execution time; they are for illustration purposes only.

Let's now investigate what "bad" looks like in its many guises. While executing a query using SnowSight, if you suspect the query has not completed within a reasonable timeframe, clicking the ID will display the available query profile.

The most important message from this section is to simplify code.

Notes on Data Capture

In this section, you will create tables and select from the Account Usage Store the `query_history` table. You can also create a view to overlay `query_history`; note that you also use a seven-day time band.

The choice of tables is deliberate. Your code base is intended to be extensible for deployment and data capture from a central provisioning team for which my previous book *Maturing the Snowflake Data Cloud* provides a detailed hands-on guide.

You can assume performance tuning is not a one-off activity; you should pro-actively monitor, detect, and continually remediate performance issues. This chapter establishes a variety of tools and techniques used in support of these activities, and the last chapter brings everything together. I can't wait to get there!

Join Explosion

Let's start with a Cartesian product, which can be considered a "join explosion." In this section, I will explain what a Cartesian join is, how to identify one, and finally offer information on how to remediate Cartesian joins.

What Is a Cartesian Join?

A Cartesian join, Cartesian product, cross join, or join explosion occurs when a join condition is missing from query predicates resulting in every combination of rows for the absent join condition being returned.

```
USE WAREHOUSE IDENTIFIER ( $tpc_warehouse_xs );
```

Suspend and restart the warehouse.

```
ALTER WAREHOUSE IDENTIFIER ( $tpc_warehouse_xs ) SUSPEND;

ALTER WAREHOUSE IDENTIFIER ( $tpc_warehouse_xs ) RESUME;
```

You can use the TPC baseline data set to illustrate a Cartesian join output. In the following example, region_baseline has five rows, and nation_baseline has 25 rows, with the join key regionkey being omitted. Assuming regionkey is matched in both tables, you should expect 25 rows. As regionkey is missing, 125 rows are returned.

```
SELECT r.r_name   AS region_name,
       n.n_name   AS nation_name
FROM   region_baseline r,
       nation_baseline n
ORDER BY n.n_name;
```

Figure 3-14 shows the corresponding query profile. Note that the CartesianJoin and row count are highlighted.

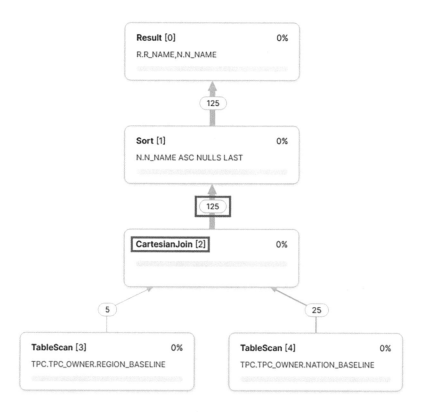

Figure 3-14. *Cartesian join query profile*

A valid, but rare, use case for a Cartesian join is when generating test data—lots of test data. Selecting from two tables without a join condition results in the number of rows returned from the first table multiplied by the number of rows in the second table.

To demonstrate a Cartesian product using EXPLAIN, let's create a suboptimal SQL statement by excluding one join condition from the previous example query used to create v_supplier_part.

```
EXPLAIN USING TABULAR
SELECT p.p_name          AS part_name,
       p.p_retailprice   AS retail_price,
       ps.ps_supplycost  AS supply_cost,
       ps.ps_availqty    AS available_quantity,
       ps.ps_supplycost / ps.ps_availqty   AS unit_price,
       s.s_acctbal       AS supplier_account_balance,
       s.s_name          AS supplier_name
```

```
FROM    partsupp_baseline ps,
        part_baseline     p,
        supplier_baseline s
WHERE   ps.ps_partkey     = p.p_partkey;
```

As every SQL developer will attest, it's easy to miss a join condition particularly where composite natural keys are used. A query profile provides the means to spot these conditions, and Figure 3-15 shows a CartesianJoin clearly indicated for our example query.

step	id	parent	operation	objects	alias	expressions
null	null	null	GlobalStats	null	null	null
1	0	null	Result	null	null	P.P_NAME, P.P_RETAILPRICE, PS.PS_SUPPLYC
1	1	0	InnerJoin	null	null	joinKey: (PS.PS_PARTKEY = P.P_PARTKEY)
1	2	1	TableScan	TPC.TPC_OWNER.PARTSUPP_BASELINE	PS	PS_PARTKEY, PS_AVAILQTY, PS_SUPPLYCOST
1	3	1	CartesianJoin	null	null	null
1	4	3	TableScan	TPC.TPC_OWNER.SUPPLIER_BASELINE	S	S_NAME, S_ACCTBAL
1	5	3	JoinFilter	null	null	joinKey: (PS.PS_PARTKEY = P.P_PARTKEY)
1	6	5	TableScan	TPC.TPC_OWNER.PART_BASELINE	P	P_PARTKEY, P_NAME, P_RETAILPRICE

Figure 3-15. *Cartesian join tabular output*

As previously stated, profiling every SQL statement as part of our continuous integration testing and then scanning the output for the CartesianJoin keyword would capture this particular problem before delivery.

Identifying Cartesian Joins

In a production system you might not immediately know which query is causing a Cartesian join, particularly after a new software release where a robust continuous integration practice has not been implemented. In this scenario, you are looking to identify rogue queries from all of those queries that have been executed.

Let's first create a table to hold all candidate query IDs for later investigation. I leave it to you to add timestamps, etc.

```
CREATE OR REPLACE TABLE cartesian_join_queries
(
sp_name                 STRING,
query_id                STRING,
```

```
operator_type         STRING,
operator_id           NUMBER,
operator_attributes   VARIANT,
row_multiple          NUMBER
);
```

To work around the GET_QUERY_OPERATOR_STATS single-input value limitation, you will create a JavaScript stored procedure called sp_get_cartesian_join_queries.

Snowflake scripting can also be used to deliver equivalent functionality and may be preferrable; you can find further information at https://docs.snowflake.com/en/developer-guide/stored-procedure/stored-procedures-snowflake-scripting. I leave it to you to convert the example code from JavaScript to SQL scripting.

Regardless of implementation approach, you may want to amend the core query driving predicates.

```
CREATE OR REPLACE PROCEDURE sp_get_cartesian_join_queries()
RETURNS string
LANGUAGE javascript
EXECUTE AS CALLER
AS
$$
   var sql_stmt  = "";
   var err_state = "";
   var recset    = "";
   var query_id  = "";
   var result    = "";

   sql_stmt  = "SELECT query_id\n"
   sql_stmt += "FROM    snowflake.account_usage.query_history\n"
   sql_stmt += "WHERE   query_type      IN ( 'SELECT', 'CREATE_TABLE_AS_
   SELECT' )\n"
   sql_stmt += "AND     warehouse_name IS NOT NULL\n"
   //Exclude cached results
   sql_stmt += "AND     execution_status   = 'SUCCESS'\n"
   //Include completed queries
   sql_stmt += "AND     bytes_scanned      > 0\n"
   //Must have scanned data
```

```
sql_stmt += "AND    total_elapsed_time > 1000;"
//Execution time must be over 1s, queries which used compute

stmt = snowflake.createStatement( { sqlText: sql_stmt } );

try
{
   recset = stmt.execute();
   while(recset.next())
   {
      query_id  = recset.getColumnValue(1);
      sql_stmt  = "INSERT INTO cartesian_join_queries\n"
      sql_stmt += "SELECT 'sp_get_cartesian_join_queries',\n"
      sql_stmt += "        query_id,\n"
      sql_stmt += "        operator_type,\n"
      sql_stmt += "        operator_id,\n"
      sql_stmt += "        operator_attributes,\n"
      sql_stmt += "        operator_statistics:output_rows /\n"
      sql_stmt += "          operator_statistics:input_rows AS row_
      multiple\n"
      sql_stmt += "FROM    TABLE ( get_query_operator_stats('" + query_id
      + "'))\n"
      sql_stmt += "WHERE  operator_type = 'CartesianJoin';"
      stmt = snowflake.createStatement ({ sqlText:sql_stmt });
      try
      {
         stmt.execute();
         result = "Success";
      }
      catch { result = sql_stmt; }
   }
   result = "Success";
}
catch(err)
{
   err_state += "\nFail Code: " + err.code;
   err_state += "\nState: " + err.state;
```

```
      err_state += "\nMessage : " + err.message;
      err_state += "\nStack Trace:\n" + err.StackTraceTxt;
      result = err_state;
   }

   return result;
$$;
```

With the stored procedure created, you now invoke it with the following:

```
CALL sp_get_cartesian_join_queries();
```

Finally, examine the result sets but query the `cartesian_join_queries` table.

```
SELECT query_id,
       operator_type,
       operator_id,
       operator_attributes,
       row_multiple
FROM   cartesian_join_queries;
```

Your results will vary according to the workload carried out before testing. There is one additional point to note: the `row_multiple` value is always one less than the value expected, because the first value represents the original row, and all other values are duplicates.

Cartesian Join and Join Explosion Costs

There are significant impacts from Cartesian joins; we list some of them here:

- Significantly larger data sets than might be expected are returned.

- The query runtime is excessively long.

- They cause spills to disk; I will discuss this later.

- They consume expensive warehouse resources.

- They cost you money.

Remediating Cartesian Joins and Join Explosions

Now that you understand what a Cartesian join is, how to identify one, and the costs associated with them, you can turn your attention to remediating identified queries.

The logical answer is to identify missing join criteria as this is the most likely root cause. Note that missing composite key attributes are far harder to identify than single attribute primary key/foreign key relationships. A general rule of thumb is that the number of <u>WHERE</u>/<u>AND</u> join conditions should always equal the number of tables minus 1. This works for many scenarios.

Nonunique key joins may also produce partial Cartesian joins. To resolve this issue, deconstruct your query into its constituent parts and check that each part returns the expected cardinality. Equality joins most often result in faster processing speed than nonequality joins, which should be avoided.

Numeric data type joins are the fastest of all. I prefer sequence generated surrogate primary keys over natural or composite keys for all tables along with declared referential integrity.

The Snowflake optimizer prefers conjunctive (additive) joins. These are predicates with <u>AND</u> operators; predicates with <u>OR</u> operators are disjunctive (subtractive) joins that are known to affect performance. Disjunctive joins should be rewritten using UNION/UNION ALL to improve performance.

The table join order can also be significant. Start with the smallest tables first as this may eliminate the greatest number of micro-partitions early within the query optimization stage. Also check that the filter criteria are sufficiently selective to improve micro-partition pruning.

Ultimately, consider refactoring the query to eliminate bottlenecks, which may include using temporary tables as intermediary storage for the large result sets.

Long Compilation Time

Within this section I define long compilation time, provide the means to identify queries suffering from long compilation time, and then finally offer remediation steps to resolve this issue.

What Is Long Compilation Time?

Any query can be considered to have a long compilation time where the query compilation time exceeds the query execution time. In other words, more time is spent compiling a query than performing real work in executing and delivering the result set.

As you know, compute is expensive. Our objective is to squeeze the maximum performance from the system. Therefore, understanding the causes of long compilation time is important, after which you may be able to remediate the root cause.

Sometimes, long compilation time is unavoidable and works as expected.

Identifying Long Compilation Time

In a production system, you might not immediately identify queries suffering from long compilation time. You are looking to identify these queries from all of those queries that have been executed.

Let's first create a table to hold all candidate query IDs for later investigation. I leave it to you to add timestamps, etc.

```
CREATE OR REPLACE TABLE long_compilation_time
(
query_id            STRING,
warehouse_name      STRING,
warehouse_size      STRING,
compilation_time_ms NUMBER,
execution_time_ms   NUMBER,
time_multiple       NUMBER  //Compilation / Execution
);
```

Latency for QUERY_HISTORY may be up to 45 minutes.

You can now insert candidate records from QUERY_HISTORY. Note that you only want records where the compilation time exceeds the execution time.

```
INSERT INTO long_compilation_time
SELECT query_id,
       warehouse_name,
       warehouse_size,
       compilation_time,
       CASE execution_time
           WHEN 0 THEN 1
           ELSE execution_time
       END AS execution_time_1,
       compilation_time / execution_time_1
FROM   snowflake.account_usage.query_history
WHERE  ( compilation_time / execution_time_1 ) > 1;
```

You can find further information on QUERY_HISTORY at https://docs.snowflake.com/en/sql-reference/account-usage/query_history.

You can examine summary information for queries where the compilation time exceeds the execution time.

```
SELECT query_id,
       warehouse_name,
       warehouse_size,
       compilation_time_ms,
       execution_time_ms,
       time_multiple
FROM   long_compilation_time;
```

From the previous result set, use GET_QUERY_OPERATOR_STATS to examine statistics for individual operators within a single query.

```
SELECT *
FROM   TABLE ( get_query_operator_stats('<your query_id here>'));
```

I leave it to you to refine the previous queries.

Long compilation time monitoring for trends is worth considering as this metric may be a leading indicator of future performance issues.

Long Compilation Time Costs

Long compilation times may be caused by many factors; I list some here:

- High data volumes in tables

- Micro-partition fragmentation due to a high number of low-volume INSERTs and UPDATEs

- Highly denormalized tables with lots of attributes

- Multiple levels of nested view decomposition required to resolve objects

- Multiple levels of role hierarchy navigation required to resolve objects

- Multiple data masking policies to resolve attribute content

- Multiple row-level access policies to resolve entitled data sets

- Number of and complexity of expressions applied to attributes

- Degree of pruning required to resolve data sets

- Lack of bind variable use negating query reuse

Remediating Long Compilation Time Queries

Now that you understand what a long compilation time is and how to identify one, you can turn your attention to remediating identified queries.

Snowflake optimally prefers 10 or fewer attributes to be returned in a result set, avoiding attribute lists of more than 100.

Wherever possible, simplify queries by reducing the number of views navigated. And for existing tables where the cluster key does not match the query predicates, consider adding a materialized view. In highly volatile environments with unpredictable workloads, I will demonstrate how to externally parallelize queries in a later chapter to reduce runtimes.

Consider using dynamic tables to offload work onto serverless compute. Note that this feature at the time of writing is in public preview, and the refresh time may be a factor in your decision-making.

Complex role hierarchies take time to navigate and reduce the number of database roles required to resolve object dependencies.

The importance of bind variables within queries is often overlooked, particularly by more junior programmers. Bind variables enable query reuse requiring a hard parse only for the first time they are seen by the query optimizer; all subsequent query submissions will reuse the original execution plan.

In contrast, while the fabric of a SQL statement may remain static, without using bind variables, the literals embedded will always force a hard-parse. As subsequent query submissions are seen as new statements, query reuse cannot occur.

Best practice is to implement bind variables where queries are to be reused. The small overhead in development cost is always returned several times in lower execution costs. Furthermore, encapsulation of reusable queries using bind variables within stored procedures is a must-have to reduce both complexity and development cost.

To illustrate the use of bind variables, you will create a JavaScript stored procedure called sp_test_bind. Note that SQL scripting can also be used.

Within sp_test_bind, two statements implement bind variables.

The first statement sql_stmt = "SELECT :1" provides a placeholder for the bind variable :1.

The second statement stmt = snowflake.createStatement({ sqlText: sql_stmt, binds:[P_NAME] }); declares the bind variable to replace :1 at runtime.

Note in JavaScript the bind variable name must be in UPPERCASE to reference parameters passed into the stored procedure.

```
CREATE OR REPLACE PROCEDURE sp_test_bind( P_NAME STRING )
RETURNS string
LANGUAGE javascript
EXECUTE AS CALLER
AS
$$
   var stmt      = "";
   var sql_stmt  = "";
   var err_state = "";
   var retval    = "";
   var result    = "";

   sql_stmt  = "SELECT :1"

   stmt = snowflake.createStatement( { sqlText: sql_stmt, binds:[P_
   NAME] } );
```

```
    try
    {
        retval = stmt.execute();
        while(retval.next())
        (
            result = retval.getColumnValue(1)
        )
    }
    catch(err)
    {
        err_state  = sql_stmt;
        err_state += "\nFail Code: " + err.code;
        err_state += "\nState: " + err.state;
        err_state += "\nMessage : " + err.message;
        err_state += "\nStack Trace:\n" + err.StackTraceTxt;
        result = err_state;
    }

    return result;
$$;
```

To call sp_test_bind, use the following CALL statement:

```
CALL sp_test_bind( 'Andrew Carruthers' );
```

Long compilation time monitoring for trends is worth considering as this metric may be a leading indicator of future performance issues. This metric can provide early warning of data skewing.

Long Execution Time

Long execution time is typically a trend-based metric where progressive or sudden adverse changes in either a single query or several queries may be observed. Monitoring long execution time may be a leading indicator of future performance issues.

What Is Long Execution Time?

Long execution time occurs after a query has been compiled and relates to the physical amount of time required to return a result set. You can use the query history total elapsed time as the sole indicator of query execution time.

Identifying Long Execution Time

In a production system, you might not immediately identify queries suffering from long execution time; individual query runtimes may be hidden within complex data pipelines where multiple SQL statements are executed sequentially. The presenting symptoms may be hidden within the overall runtime of a process or the backup of files awaiting processing. Conversely, for queries identified as long running, you need to identify which data pipeline or process they belong to before remediation can occur.

In this section, you are looking to identify long-running queries from all of the queries that have been executed.

Let's first create a table to hold all candidate query IDs for later investigation. I leave it to you to add timestamps, etc.

```
CREATE OR REPLACE TABLE long_execution_time
(
query_id                 STRING,
warehouse_name           STRING,
warehouse_size           STRING,
query_execution_time_ms  NUMBER,
partitions_scanned       NUMBER,
partitions_total         NUMBER
);
```

The latency for QUERY_HISTORY may be up to 45 minutes.

You can now insert candidate records from QUERY_HISTORY; in this example you want records only for the past seven days; you may want to change the predicates to suit your requirements.

```
INSERT INTO long_execution_time
SELECT query_id,
       warehouse_name,
       warehouse_size,
       total_elapsed_time / 1000 AS query_execution_time_ms,
       partitions_scanned,
       partitions_total
FROM   snowflake.account_usage.query_history
WHERE  cluster_number IS NOT NULL     //Exclude cached results
AND    execution_status   = 'SUCCESS' //Include completed queries
AND    bytes_scanned      > 0         //Must have scanned data
AND    total_elapsed_time > 1000      //Execution time must be over 1s,
queries which used compute
AND    TO_DATE ( start_time ) > DATEADD ( day, -7, TO_DATE ( current_
timestamp()));
```

You can find further information on QUERY_HISTORY at https://docs.snowflake.com/en/sql-reference/account-usage/query_history.

```
SELECT query_id,
       warehouse_name,
       warehouse_size,
       query_execution_time_ms,
       partitions_scanned,
       partitions_total
FROM   long_execution_time;
```

From the previous result set, use GET_QUERY_OPERATOR_STATS to examine statistics for individual operators within a single query.

```
SELECT *
FROM   TABLE ( get_query_operator_stats('<your query_id here>'));
```

I leave it to you to refine the previous queries.

Long Execution Time Costs

Long execution times may be caused by many factors; I list some here:

- Micro-partition fragmentation due to a high number of low-volume inserts and updates

- Degree of pruning required to resolve data sets

- Data skewing resulting from changing data content over time

- Query predicates not matching cluster key definition

- High cardinality queries with many selective criteria leading to inefficient pruning

Remediating Long Execution Time Queries

Automatic clustering can remediate long execution times by re-ordering micro-partition content into optimally efficient form. Note that automatic clustering can invalidate cached results. I will discuss automatic clustering in detail in the next chapter. You can find further details at `https://docs.snowflake.com/en/user-guide/tables-auto-reclustering`.

For existing tables where the cluster key does not match the query predicates, consider adding a materialized view. Consider using dynamic tables to offload work onto serverless compute. Note that the refresh time may be a factor in your decision-making.

Over time data within a table may become skewed. This occurs as data ranges change over time and is often a side effect of high `INSERT` and `UPDATE` activity where micro-partition content changes frequently, a phenomena referred to as *churn*. I will discuss this phenomena in detail within the next chapter, but for now, it is sufficient to know skewed data can impact query execution time as the number of micro-partitions scanned may be higher than expected.

A search optimization service can improve the performance of highly selective queries but is not recommended for high-churn environments. I discuss search optimization services in detail later within this book. You can find further details at `https://docs.snowflake.com/en/user-guide/search-optimization-service`.

Incorrect join order can also contribute to long execution time; put the lowest cardinality table first after the FROM clause, and put the highest cardinality table last.

Long Table Scan

Long table scans manifest as a high percentage value for the `TableScan` operator within a query profile. This may become a trend-based metric where progressive or sudden adverse changes in performance of either a single query or several queries may be observed. Monitoring long table scans may be a leading indicator of future performance issues.

What Is Long Table Scan?

A long table scan occurs where most of the processing time is spent servicing remote disk I/O, an expensive operation. I discuss local and remote disk I/O in detail within the next chapter. You will also experience long table scans where there is little or no partition pruning, identified from the query profile summary.

You might reasonably expect to see long table scans as part of your day-to-day operational processes when creating denormalized table content; therefore, local system knowledge is required when interpreting results. Alternatively, you might expect to see long table scans when exploring data sets as part of an investigation or in preparation for further system development.

Identifying Long Table Scans

In a production system you might not immediately identify queries suffering from long table scans as the metric is recorded as steps within each query profile. The presenting symptoms may be hidden within the overall runtime of a process or with backup files awaiting processing. Conversely, for queries identified as having long table scans, you need to identify which data pipeline or process they belong to before remediation can occur.

In this section you are looking to identify long table scans from all of the queries that have been executed.

Let's first create a table to hold all candidate query IDs for later investigation. I leave it to you to add timestamps, etc.

```
CREATE OR REPLACE TABLE long_table_scans
(
query_id              STRING,
warehouse_name        STRING,
warehouse_size        STRING,
```

```
partition_scan_ratio      NUMBER,
partitions_scanned        NUMBER,
partitions_total          NUMBER
);
```

The latency for QUERY_HISTORY may be up to 45 minutes.

You can now insert candidate records from QUERY_HISTORY. In this example you can identify those records with a micro-partition scanned to a total ratio greater than 50 percent; you may want to change the predicates to suit your requirements.

```
INSERT INTO long_table_scan
SELECT query_id,
       warehouse_name,
       warehouse_size,
       partitions_scanned / partitions_total AS partition_scan_ratio,
       partitions_scanned,
       partitions_total
FROM   snowflake.account_usage.query_history
WHERE  warehouse_name IS NOT NULL     //Exclude cached results
AND    execution_status   = 'SUCCESS' //Include completed queries
AND    bytes_scanned      > 0         //Must have scanned data
AND    total_elapsed_time > 1000      //Execution time must be over 1s,
                                              queries which used compute
AND    ( partitions_scanned / partitions_total ) > 0.5;
```

You can find further information on QUERY_HISTORY at https://docs.snowflake.com/en/sql-reference/account-usage/query_history.

```
SELECT query_id,
       warehouse_name,
       warehouse_size,
       partition_scan_ratio,
       partitions_scanned,
       partitions_total
FROM   long_table_scan;
```

From the previous result set, use GET_QUERY_OPERATOR_STATS to examine statistics for individual operators within a single query.

```
SELECT *
FROM    TABLE ( get_query_operator_stats('<your query_id here>'));
```

I leave it to you to refine the previous queries.

Long table scan monitoring for trends is worth considering as this metric may be a leading indicator of future performance issues. I suggest this metric can provide early warning of data skewing.

Long Table Scan Costs

Assuming you can exclude long table scans due to known and expected behaviors, the remaining long table scan candidates can be caused by many factors.

- Micro-partition fragmentation due to a high number of low-volume inserts and updates

- Query predicates not matching table cluster key

- Range scan filtering occurring when using BETWEEN, LIKE, <>, and similar operators

Remediating Long Table Scan Queries

Automatic clustering can remediate long execution times by re-ordering micro-partition content into optimally efficient form. Note that automatic clustering can invalidate cached results. I discuss automatic clustering in detail in the next chapter. You can find further details at https://docs.snowflake.com/en/user-guide/tables-auto-reclustering.

For existing tables where the cluster key does not match the query predicates, consider adding a materialized view. Consider using dynamic tables to offload work onto serverless compute. Note the refresh time may be a factor in your decision-making.

Improve query selectivity to match the cluster key definition on the target table.

Consider using a query acceleration service to dynamically scale and parallelize portions of the query plan leading to overall reduced runtime. I discuss query acceleration services in detail later within this book. You can find further details at https://docs.snowflake.com/en/user-guide/query-acceleration-service.

Spills to Disk and Out of Memory

We will consider spill to disk and out of memory (OOM) within the same section; they are related because spills can cause OOM events.

Every warehouse has a finite amount of memory allocated; you may recall from Chapter 2 that I stated an extra small (XSmall) warehouse has eight CPUs and associated cache, about 16 to 24GB RAM, local SSD storage, and remote attached storage. For every size you increase your warehouse, the number of CPUs doubles and memory increases too.

What Causes a Spill to Disk and OOM?

A spill to disk occurs when a query attempts to consume more memory than the warehouse has available for allocation. In this scenario, intermediate results are first spilled to local SSD storage and then to remote storage, before finally exceeding all available memory and storage resulting in an OOM error. Snowflake will then attempt to retry the query.

As Snowflake dynamically allocates CPU and memory, it is possible the warehouse workload has reduced between query failure and query retry, rendering more resources available for the retry attempt.

For your further investigation, an example scenario for spills to disk and OOM using `ORDER BY` and `LIMIT/OFFSET` is available at `https://community.snowflake.com/s/article/Out-of-memory-error-caused-by-LIMIT-and-or-OFFSET-clause`.

You must understand the importance of sizing warehouses correctly according to the expected workload. You also see how queries returning large volumes of data with smaller warehouses may exceed the available memory and cause an OOM failure.

Identifying Spills to Disk

Spills to disk are readily identified from the statistics block of every query profile where "spilling" information is presented when spills to disk occur.

In this section, you will identify both spills to disk and OOMs from all of those queries that have been executed.

To illustrate a spill, let's deliberately incorrectly size a warehouse and then execute a poor-quality query. Note that you should never use `SELECT  *` for your production code.

```
USE WAREHOUSE IDENTIFIER ( $tpc_warehouse_xs );

SELECT *
FROM    partsupp_baseline ps,
        part_baseline     p,
        supplier_baseline s
WHERE   ps.ps_partkey     = p.p_partkey
AND     ps.ps_suppkey     = s.s_suppkey;
```

You do not need a poor-quality query to complete before viewing spills, which are visible via the Profile Overview tab accessible by clicking the query ID. Spills to disk are indicated by both Local Disk I/O and Remote Disk I/O, as shown in Figure 3-16.

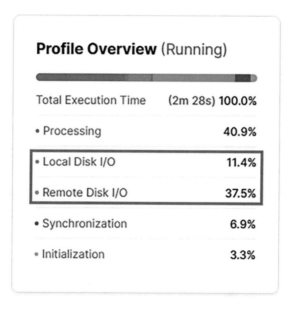

Figure 3-16. *Spills to disk*

Having identified a spill, let's create a table to hold all the candidate query IDs for later investigation. I leave it to you to add timestamps, etc.

```
CREATE OR REPLACE TABLE spill_and_OOM
(
query_id                        STRING,
warehouse_name                  STRING,
warehouse_size                  STRING,
```

```
bytes_spilled_to_local_storage    NUMBER,
bytes_spilled_to_remote_storage   NUMBER,
bytes_sent_over_the_network        NUMBER
);
```

The latency for QUERY_HISTORY may be up to 45 minutes.

You can now insert the candidate records from QUERY_HISTORY; in this example, you will identify those records where both local and remote spills to storage have occurred. In other words, both values are greater than zero. You may want to change the predicates to suit your requirements.

```
INSERT INTO spill_and_OOM
SELECT query_id,
       warehouse_name,
       warehouse_size,
       bytes_spilled_to_local_storage,
       bytes_spilled_to_remote_storage,
       bytes_sent_over_the_network
FROM   snowflake.account_usage.query_history
WHERE  warehouse_name IS NOT NULL     //Exclude cached results
AND    bytes_spilled_to_local_storage  > 0;
```

You can find more information on QUERY_HISTORY at https://docs.snowflake.com/en/sql-reference/account-usage/query_history.

You can examine summary information for queries where spills and potential OOMs have occurred.

```
SELECT query_id,
       warehouse_name,
       warehouse_size,
       bytes_spilled_to_local_storage,
       bytes_spilled_to_remote_storage,
       bytes_sent_over_the_network
FROM   spill_and_OOM;
```

From the previous result set, use GET_QUERY_OPERATOR_STATS to examine statistics for individual operators within a single query.

```
SELECT *
FROM    TABLE ( get_query_operator_stats('<your query_id here>'));
```

I will leave it to you to refine the previous queries.

I highly recommend monitoring for spills as this metric is a leading indicator of future problems and may identify remediation opportunities well before OOMs occur.

Spills to Disk and OOM Costs

Earlier in this section I explained the root cause of spills to disk and OOMs; I now list some additional factors here:

- High workload concurrency where the warehouse cannot service all queries at the same time

- Unexpectedly high volumes of data processed

- Incorrectly sized warehouse for workload

- Large intermediate result sets

Remediating Spills to Disk and OOM Queries

Automatic clustering can remediate long execution times by re-ordering micro-partition content into an optimally efficient form. Note that automatic clustering can invalidate cached results. I will discuss automatic clustering in detail within the next chapter. You can find further details at https://docs.snowflake.com/en/user-guide/tables-auto-reclustering.

Reducing warehouse concurrency by segregating workloads into separate warehouses may free sufficient resources to remediate disk spills. Alternatively, increasing warehouse size will increase available resources, enabling more optimal in-memory operations, and reduce both local and remote spills to disk.

The table join order can also be significant. Start with the smallest tables first as this may eliminate the greatest number of micro-partitions early within the query optimization stage and then in cardinality order up to the highest cardinality last. Also check the filter criteria are sufficiently selective to improve micro-partition pruning.

Ultimately, consider refactoring the query to eliminate bottlenecks, which may include using temporary tables as the intermediary storage for large result sets.

Join Order

Table join order has the capability to derail the best of queries. As with everything, a little knowledge can be dangerous, and applying expert knowledge from legacy RDBMSs is particularly dangerous.

I suggest an open mind when investigating query performance issues.

In Snowflake, the table placement order is opposite to what you would expect in Oracle.

Why Is Join Order Important?

Referring to our earlier explanation of what a "good" query profile looks like, when the build side is larger than the probe side, performance is usually slower.

You also know Snowflake builds hash tables in preparation for joining data sets, and the order in which you join your tables has significance. For all query profiles, and assuming a right deep join tree, the optimal pattern is for the table returning the largest data set at the bottom right and the table returning the smallest data set at the top left. You saw an example of a right deep join tree earlier within this chapter.

Within your SQL statement, you should place the table with the lowest cardinality first within the FROM clause as this has the potential to prune the most micro-partitions earliest within the query. The next lowest cardinality table should be second, and so on, up to the highest cardinality table last.

Identifying Join Order Issues

When investigating join order issues, you should first examine the query profile. As you can see from Figure 3-17, the optimal pattern for result set determination is the largest data volume at the bottom right, with reduction to the smallest data volume shown at the top left.

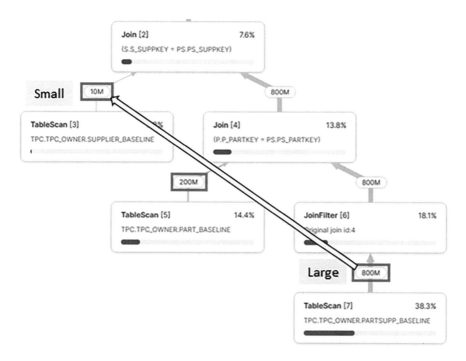

Figure 3-17. Optimal result set pattern

Needless to say, most query profiles do not conform to the pattern shown in Figure 3-17, but it is helpful to understand what "good" looks like and then iterate toward the optimal query profile as you tune your queries.

You might also experience join order issues when missing attributes from natural key joins or missing primary key/foreign key relationships. Both of these scenarios result in join explosions.

Detection of join order issues is largely through examination of individual queries and visual identification and then testing data volumes for each referenced table. You may want to consider using a clone of your production environment to determine representative results.

Poor Join Order Costs

Poor join criteria and ordering can result in the following:

- Inefficient joins where the build side hash tables are larger than optimal resulting in higher probe execution times.

- Join explosion resulting in spillage and OOMs due to large intermediate data set generation.

- Left deep tree join refers to query profiles where the predominating query profile branches to the left, the opposite of a right deep tree join. Left deep tree joins consume more warehouse memory and reduce parallel processing options.

Remediating Poor Join Order Issues

A general rule of thumb is that the number of WHERE/AND join conditions should always equal the number of tables minus 1. This works for many scenarios except for composite natural keys.

You should also be mindful that the Snowflake query optimizer prefers numeric data type joins. These join criteria can also cause performance issues:

- **Data type conversion:** Joining a NUMBER to a VARCHAR.

- **Evaluate expressions:** MIN/MAX, etc.

- **User-defined functions (UDFs):** Complex logic embedded into a UDF.

- **Common table expression (CTE):** You will investigate CTEs later in this chapter.

We have all encountered complex SQL statements that are hard to read. On closer examination you may find the following:

- SELECT *

- Redundant table joins

- Missing composite key join attributes

- DISTINCT forcing uniqueness

- Unnecessary join keys

Remediating poor join order issues involves a lot of time and hard work along with constant retesting of changes with the eventual objective of improving performance. You must at all times simplify your code wherever possible by removing complexity.

While automatic clustering and search optimization can help, automatic clustering can invalidate cached results. There is no substitute for well-formed and optimally performing SQL statements.

Common Table Expressions

This chapter was inspired by the post at `https://select.dev/posts/should-you-use-ctes-in-snowflake` by Niall Woodward (@NiallWoodward).

CTEs are individually named temporary result sets built within the SQL statement. A SQL statement may have none, one, or many CTEs. They are identified by the presence of a `WITH` clause before the `SELECT` statement. CTEs may also be referred to as *subquery refactoring* and are supported by many RDBMSs in addition to Snowflake.

The following is the general form of a CTE:

```
WITH <subquery>
SELECT <attributes>
FROM    <table>
WHERE   <predicates>
ORDER BY <ordering>;
```

Simple CTE Use Case

CTEs are often used to create subsets of data that may be repeatedly used within the main body of the `FROM/WHERE` clause to simplify code. A common use case is to use `UNION ALL` to join two tables into a single CTE thus simplifying the main body of code, as this next example shows:

```
USE WAREHOUSE IDENTIFIER ( $tpc_warehouse_xs );
```

Create an example SQL statement with a CTE:

```
WITH regiongroup AS
(
SELECT r_regionkey,
       r_name,
       'EMEA'  AS r_regiongroup
FROM   region_baseline
```

```
WHERE   r_name IN ( 'EUROPE', 'MIDDLE EAST' )
UNION ALL
SELECT r_regionkey,
       r_name,
       'APAC'  AS r_regiongroup
FROM    region_baseline
WHERE   r_name = 'ASIA'
)
SELECT rg.r_regiongroup AS region_group,
       rg.r_name        AS region_name,
       n.n_name         AS country_name
FROM    regiongroup      rg,
        nation_baseline  n
WHERE   rg.r_regionkey   = n.n_regionkey
ORDER BY 1, 2, 3;
```

When you examine the resultant query profile, you can see that the output of the build side (UNION ALL output) has been pushed down to the probe side as evidenced by the row counts returned from the TableScan(9), as shown in Figure 3-18.

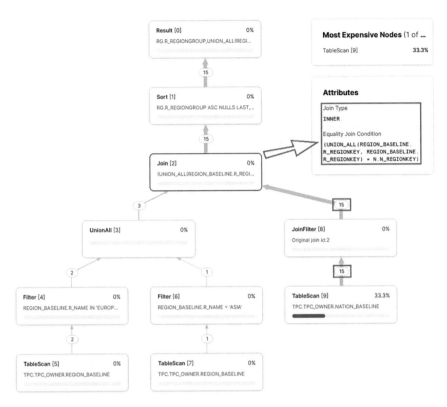

Figure 3-18. *Filter push down*

However, there are scenarios where CTEs may not perform as expected, which I will discuss next.

Reusing CTEs

Where a CTE is referenced more than twice within the same SQL statement, you may find attribute pruning is disabled.

In this example, a query used to expose the CTE reuse issue; you can declare a parent CTE that is referenced by two child CTEs. Each child is then referenced by the main query body on either side of a UNION ALL.

```
WITH nation_list AS
(
SELECT r.*
FROM   nation_baseline r
),
```

```
comment_list AS
(
SELECT n_comment
FROM    nation_list
),
name_list AS
(
SELECT n_name
FROM    nation_list
)
SELECT n_comment
FROM    comment_list
UNION ALL
SELECT n_name
FROM    name_list;
```

You might reasonably expect only the unique attributes to be SELECTed from the nation_baseline table, but as the query profile shown in Figure 3-19, you also see the highlighted attribute N_REGIONKEY, which is not referenced in any CTE or the main body.

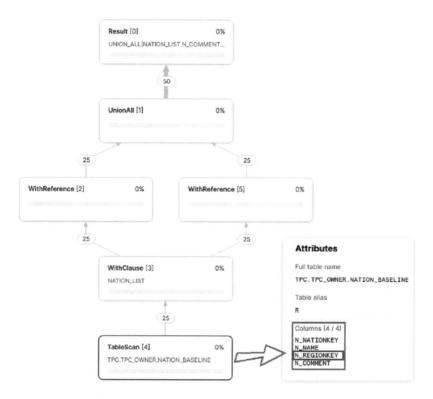

Figure 3-19. *CTE attribute pushdown disabled*

The important point to note is the single table scan TableScan2 showing the CTE is resolved once and no filters are pushed down.

You will not experience the same behavior when explicitly referencing the same base table nation_baseline as used within the earlier parent CTE.

```
WITH comment_list AS
(
SELECT n_comment
FROM   nation_baseline
),
name_list AS
(
SELECT n_name
FROM   nation_baseline
)
SELECT n_comment
```

```
FROM    comment_list
UNION ALL
SELECT n_name
FROM    name_list;
```

Not only is the refactored query easier to read, the query profiler is also simpler, as shown in Figure 3-20 where the predicate pushdown is highlighted.

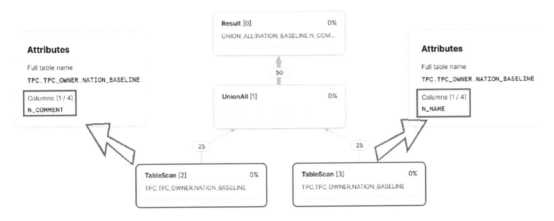

Figure 3-20.   *CTE attribute pushdown enabled*

CTE Costs

I recommend the use of CTEs to abstract complex logic and simplify code, but not in all situations. As you have seen, nested CTEs can lead to predicates not being pushed down.

Elegant code is both readable and readily understood. CTEs in general aid readability, but if poorly structured and implemented, they can increase code complexity and maintenance overheads. You should also appreciate the lost opportunity cost of developers understanding before they begin to refactor or remediate code. In other words, Keep It Simple, Stupid (KISS). Use CTEs judiciously to reduce complexity and layers of code, not forgetting to comment your code.

Remediating CTEs

Here are some tips:

- Replace CTEs with views or direct embedding within the core SQL statement.

- Simplify CTEs by removing complexity, nesting, layering code, and using recursive calls.

- Replace SELECT * with explicit attribute names.

- Denormalize data structures using dynamic tables to remove dependency on CTEs. Note that dynamic tables at the time of writing are still in in public preview.

- Refactor data pipelines to deliver denormalized tables as part of the process and not as an afterthought. Tune the design!

Summary

You began this chapter by creating a database, schema, and role to begin your investigations into query profiles with the express intent of reusing your new environment throughout the remainder of this book. You also declared warehouses sized between X-Small and X-Large and very briefly investigated the effect of using different sized warehouses on query performance.

Having identified where to access query profiles, you used the new environment to create an example query to fulfil an imaginary business requirement. Using the example query as a starting point, you began investigating query profiles.

Optimizing query profiles is dependent upon understanding what both "good" and "bad" query profiles look like. I introduced new terminology, explained how to interpret query profiles, and showcased how to identify and remediate problems.

With a firm understanding of query profiles, in the next chapter you will investigate micro-partitions.

CHAPTER 4

Micro-partitions

You may recall from previous chapters that I discussed micro-partition pruning, but I did not explain what micro-partitions are, as they deserve a chapter of their own. This is that chapter.

At first sight, micro-partitions appear to be a simple subject to discuss; they are not. As you are about to discover, micro-partitions rapidly increases the level of complexity of the discussion. There are loose ends and references that resolve themselves in later chapters (please forgive the interdependencies; they are an unfortunate consequence of attempting to tell each story as simply as possible, while piquing your interest in later chapters). This chapter reveals the lengths Snowflake has undertaken to hide the underlying complexity of micro-partitions. You are about to scratch the surface and take a peek.

In Chapter 2, I discussed query optimization in detail; the query optimizer seeks to reduce the cost of queries by determining the optimal execution path. In this chapter, I discuss the micro-partition features to both derive optimal data management and control storage costs.

Performance tuning must consider both execution runtime and storage costs, because both aspects are inextricably related.

I will begin by explaining some foundational information that many of us who have been involved in information technology take for granted. Recent conversations with junior colleagues indicate an absence of basic information, so this section is intended to close those knowledge gaps.

The next chapter will cover cluster keys, micro-partition pruning, and optimal design patterns supporting Snowflake best practices for data warehousing.

In the book *Building the Snowflake Data Cloud*, I discussed both micro-partitions and query optimizer basics. The micro-partition story has not changed since I wrote that

A. Carruthers, *Tuning the Snowflake Data Cloud*, https://doi.org/10.1007/979-8-8688-0379-6_4

book, and some of the content presented here will therefore be familiar. However, with a deeper understanding of optimizer behavior and subsequent hands-on experience, I will offer new insights into how micro-partitions affect both performance and storage cost.

The `query_id` values used throughout this book will vary when parent SQL statements are executed against your Snowflake account.

For those with a legacy RDBMS background, I should make clear that enforced constraints are absent within Snowflake standard tables. By default, Snowflake allows constraints, primary keys, and foreign keys to be declared. Note that they inform the query optimizer, but the only enforced constraint is `NOT NULL`.

I am aware that some legacy RDBMSs use primary key tracking or change data capture to implement data distribution. The absence of enforced primary keys precludes a similar approach for Snowflake where immutable micro-partitions implement data distribution capabilities.

However, the forthcoming Unistore and hybrid tables change the Snowflake approach, at least for hybrid tables. The closest comparable feature to a primary key is a cluster key, which I will discuss in the next chapter.

There is a lot of information to cover, so let's start investigating micro-partitions!

Setup

First you will declare the session variables used throughout this chapter. Note that you may need to rerun these declarations when your browser session is opened again.

```
SET tpc_owner_role   = 'tpc_owner_role';
SET tpc_warehouse_XS = 'tpc_wh_xsmall';
SET tpc_warehouse_S  = 'tpc_wh_small';
SET tpc_warehouse_M  = 'tpc_wh_medium';
SET tpc_warehouse_L  = 'tpc_wh_large';
SET tpc_warehouse_XL = 'tpc_wh_xlarge';
SET tpc_database     = 'tpc';
SET tpc_owner_schema = 'tpc.tpc_owner';
```

With your session variables declared, you now declare your environment.

```
USE ROLE       IDENTIFIER ( $tpc_owner_role   );
USE DATABASE   IDENTIFIER ( $tpc_database     );
USE SCHEMA     IDENTIFIER ( $tpc_owner_schema );
USE WAREHOUSE  IDENTIFIER ( $tpc_warehouse_xs );
```

Session variables may appear cumbersome, but they provide a level of abstraction, and I encourage their use.

Foundational Information

In this section, I describe the basic concepts relating to storage and micro-partitions to establish some baseline information that this chapter later relies upon. I do not deep dive into foundational information but instead provide a brief summary and links to both documentation and other articles for your further investigation. I assume you have some familiarity with the basic principles of persisting data and later will develop an understanding of storage-related costs.

Centralized Storage

Regardless of whether Snowflake is deployed on AWS, Azure, or GCP, every time you create a persistent Snowflake database object such as a table, storage is automatically allocated from within the associated Snowflake VPC storage.

For reference I list the storage services provided by each cloud service provider (CSP) here:

- **AWS:** AWS Amazon Simple Storage Service (Amazon S3)

- **Azure:** Blob Storage

- **GCP:** Google Cloud Storage

S3-compatible storage is discussed later in the book.

You can find more information on the data life cycle at `https://docs.snowflake.com/en/user-guide/data-lifecycle`.

Regardless of the cloud provider, Snowflake manages all storage interactions for data warehouse core operations transparently through SQL. There are four types of storage available within Snowflake that are referred to as *stages*.

- **External**
    - Hosted on any of the three supported CSPs
    - Hosted on any of the S3-compatible storage providers
- **Named or Internal:** Hosted within the Snowflake VPC for the account CSP
- **Table:** Associated with a named table
- **User:** For internal Snowflake use only

You can find more information on the data life cycle at `https://docs.snowflake.com/en/sql-reference/sql/create-stage`.

Provisioning Snowflake on the CSP infrastructure ensures you always have enough storage available and immediate access to more for scalability.

Direct Storage Access

Direct access to storage on supported CSP external devices is possible by configuring a `STAGE` to point at uncontrolled and unaudited storage thus presenting an opportunity for data breaches and worse.

We strongly recommend `STAGE` definitions are checked to ensure direct storage mappings *do not* exist within your environment.

Best practice is to restrict `STAGE` mapping to storage via a predefined `STORAGE INTEGRATION` restricting data ingress and egress to known locations.

You set this control at the Snowflake account level:

```
USE ROLE accountadmin;
ALTER ACCOUNT SET require_storage_integration_for_stage_creation  = TRUE;
ALTER ACCOUNT SET require_storage_integration_for_stage_operation = TRUE;
```

How to implement STORAGE INTEGRATION and STAGE is not discussed further in this chapter; I cover this topic in depth within *Building the Snowflake Data Cloud*. I discuss S3-compatible storage later in this chapter as this subject has not been addressed elsewhere.

Storage Costs

Using storage incurs cost; you pay for everything you consume. Snowflake does not make a margin on storage charges but instead passes through storage charges from the cloud provider according to the region and cloud provider. Nominally for AWS, and depending upon the region, it's approximately $23/terabyte. Note that this figure does vary.

Additional storage changes are incurred when using the Time Travel and Fail-Safe features.

I do not consider the cost of maintaining micro-partitions in this section but instead focus upon the true storage cost.

You can find more information on the storage costs at https://docs.snowflake. com/en/user-guide/cost-exploring-data-storage.

Block Devices

Data storage devices such as disk drives and NAND flash memory arrange data in contiguous blocks, that is, data "chunks," and are stored sequentially in storage. Field Programmable Gate Arrays (FPGAs) may also manage storage and data access.

In the old days, disk density was referenced in terms of partitions, segments, formatting, sectors, and tracks with much effort expended to optimize expensive disk storage. You might do the same today by de-fragmenting your local PC hard disk to move disk blocks into contiguous segments, which speeds up file access as the disk read head moves only to the start of the file and not across different locations to access file segments.

I mention block devices because the manner in which storage is accessed has some parallels within Snowflake; I will discuss these shortly.

Database and Table Storage

Minimizing storage costs directly relates to performance tuning, where you can meet your data resiliency objectives by choosing optimal storage types. The hidden costs can rapidly escalate.

You must tune your database designs from the outset, and this section delivers some tools to identify storage costs. You will begin by developing a view called v_table_storage_metrics to support a wider suite of attribute reporting used later in this chapter.

```
USE ROLE IDENTIFIER ( $tpc_owner_role );

CREATE OR REPLACE VIEW v_table_storage_metrics
AS
SELECT table_catalog||'.'||
       table_schema||'.'||
       table_name              AS path_to_object,
       active_bytes /1024/1024 AS active_MB,
       active_bytes
           /1024/1024/1024     AS active_GB,
       active_bytes
           /1024/1024/1024/1024 AS active_TB,
       time_travel_bytes
           /1024/1024/1024     AS time_travel_GB,
       failsafe_bytes
           /1024/1024/1024     AS failsafe_GB,
       retained_for_clone_bytes
           /1024/1024/1024     AS retained_for_clone_GB,
       clone_group_id,
       is_transient,
       deleted
FROM   snowflake.account_usage.table_storage_metrics;
```

You can find further information at https://docs.snowflake.com/en/sql-reference/account-usage/tables. Note that all Account Usage Store views experience data latency of between 45 minutes and 3 hours; therefore, you may not immediately see the expected results.

You should also be aware of the summary database-level information for the past year, which can be found within the Account Usage Store as this next SQL statement illustrates:

```
SELECT usage_date,
       database_name                            AS db_name,
       average_database_bytes /1024/1024/1024      AS avg_db_GB,
       average_database_bytes /1024/1024/1024/1024 AS avg_db_TB,
       average_failsafe_bytes /1024/1024/1024      AS avg_fs_GB,
       average_failsafe_bytes /1024/1024/1024/1024 AS avg_fs_TB,
       deleted
FROM   snowflake.account_usage.database_storage_usage_history
ORDER BY usage_date DESC;
```

Figure 4-1 shows some sample output for my TPC database; note that this view has up to three hours latency.

USAGE_DATE	DB_NAME	AVG_DB_GB	AVG_DB_TB	AVG_FS_GB	AVG_FS_TB	DELETED
2023-09-01	TPC	411.011870861	0.4013787801	0	0	null
2023-08-31	TPC	411.011870861	0.4013787801	0	0	null
2023-08-30	TPC	411.011870861	0.4013787801	0	0	null
2023-08-29	TPC	411.011870861	0.4013787801	0	0	null
2023-08-28	TPC	411.011870861	0.4013787801	0	0	null

Figure 4-1. *Database average storage consumption*

You can find more information on the Account Usage database_storage_usage_history view at https://docs.snowflake.com/en/sql-reference/account-usage/database_storage_usage_history.

Snowflake also supplies a table function referencing the information_schema, useful for live, point-in-time investigations as information_schema views hold data for only 14 days. You can find further information at https://docs.snowflake.com/en/sql-reference/functions/database_storage_usage_history. I leave this for your later investigation.

Stages

Both internal and external stages consume storage and contribute to costs:

- Internal stages consume storage within the Snowflake VPC.

- External stages consume storage on accessible CSP or S3 compatible storage.

Storage costs accrue regardless of where storage is declared, and you should consider stage storage costs to reduce your overall storage consumption costs.

Dropping an external stage does not automatically remove files stored within the mapped location.

To identify active stages for your Snowflake account, you use the Account Usage Store STAGES view. Note that a latency of up to two hours applies. Here you create a view called v_stage_locations for ease of use.

```
CREATE OR REPLACE VIEW v_stage_locations
AS
SELECT stage_catalog||'.'||
       stage_schema||'.'||
       stage_name              AS path_to_stage,
       stage_url,
       stage_owner
FROM   snowflake.account_usage.stages
WHERE  stage_owner IS NOT NULL
ORDER BY path_to_stage;
```

You might also use the equivalent information_schema view. Note the 14-day data limitation and that each database has an information_schema; therefore, every database would require separate investigation. You can find further information on STAGES at https://docs.snowflake.com/en/sql-reference/account-usage/stages.

With your stages identified and for internal stages only, you can identify the average daily storage usage. The new view called v_stage_avg_storage provides summary information.

```
CREATE OR REPLACE VIEW v_stage_avg_storage
AS
SELECT usage_date,
       average_stage_bytes / 1024 / 1024        AS avg_stage_MB,
       average_stage_bytes / 1024 / 1024 / 1024 AS avg_stage_GB
FROM   snowflake.account_usage.stage_storage_usage_history
ORDER BY usage_date DESC;
```

As with all Account Usage Store views, latency applies, in this case of up to 120 minutes. You can find more information at `https://docs.snowflake.com/en/sql-reference/account-usage/stage_storage_usage_history`.

External storage consumption can be tracked using CSP tooling and will be charged separately than the Snowflake storage charges.

Micro-partition Overview

Now I will discuss "how" storage is managed, always keeping in mind the manner in which the Snowflake optimizer processes your SQL statements to deliver highly performant queries.

What Are Micro-partitions?

Micro-partitions are the fundamental units of storage that comprise physical tables and are immutable. Snowflake does not add, change, or remove data from an existing micro-partition. Every data change is recorded by creating new micro-partitions, and the old micro-partitions age out according to the Time Travel setting and Fail-Safe, both of which are explained in my first book, *Building the Snowflake Data Cloud*.

Unlike some legacy RDBMSs, micro-partitions do not require the periodic gathering of statistics. Snowflake guarantees statistics are always maintained for every object. To illustrate this point, consider how the results for your next SQL statement are generated.

```
SELECT count(1)
FROM   lineitem_baseline;
```

This query should return 5,999,989,709 rows, though the interesting information is "how" the row count was derived. To reveal the information source, click the query ID to access the query profile. Figure 4-2 shows the row count was derived from metadata, and the profile shows "Other" indicating that the stored statistics were referenced.

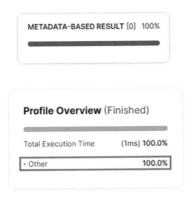

Figure 4-2. Metadata query result resolution

For Snowflake native tables, traditional indexes are not supported; therefore, index maintenance is no longer an issue. Note that hybrid tables and/or Unistore support traditional indexes and referential integrity, which is out of scope for this book but a point to remember for the future.

Immutable Micro-partitions

Micro-partition immutability offers many great benefits with few downsides. Figure 4-3 illustrates the creation of two new micro-partitions where DML activity has modified the "EMEA" contents of two micro-partitions for an imaginary table.

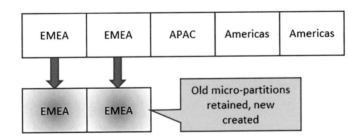

Figure 4-3. Micro-partition retention

Dependent upon table type, the retained micro-partitions may be saved for the duration of your Time Travel setting and seven-day Fail-Safe period before being permanently and irretrievably removed. You can find more information on table types, permissible Time Travel settings, and Fail-Safe periods at https://docs.snowflake. com/en/user-guide/tables-temp-transient#comparison-of-table-types.

Assuming a 90-day Time Travel retention period, you should expect your storage costs to increase by roughly 10 to 20 percent. Note that your DML profile may cause this ballpark figure to be wildly exceeded. As always, test before implementing Time Travel.

The type of table you choose will affect your storage costs for Time Travel and Fail-Safe.

Retained micro-partitions are used for the following:

- Time Travel
- Fail-Safe
- Secure Direct Data Shares
- Private Listings
- Snowflake Marketplace
- Replication
- Cloning
- Disaster recovery

I cover these subjects in great detail in *Maturing the Snowflake Data Cloud*.

Immutable micro-partitions do have some downsides. High-frequency, low-volume DML operations affecting multiple micro-partitions will result in micro-partition churn and increased data storage costs. You can also experience object locking and SQL statement queueing, which are clear indicators of performance bottlenecks.

In an extreme example, I recently experienced a table containing 2.5-billion rows having all micro-partitions rewritten over a four-hour period; hence, the Time Travel setting was one day representing the minimum period possible. Depending on your requirements, Snowflake has further provisioned an Automatic Clustering service, which may mitigate against micro-partition fragmentation, which I discuss later in this chapter.

Micro-partition Metadata

Snowflake stores data in an internal, compressed, columnar format. You also know from Chapter 1 that Snowflake captures and maintains the following statistics for each micro-partition stored in the Cloud Services layer:

- Table and micro-partition
    - Row count
    - Size in bytes (including compression information)
    - File reference
    - Table version
- Clustering
    - Total number of micro-partitions
    - Micro-partition overlap values
    - Micro-partition depth
- Column
    - Max/min value range
    - The number of distinct values
    - NULL count
- Subcolumn
    - Statistics for common paths in semi-structured data

Referencing the metadata held for each micro-partition, the optimizer is able to rapidly identify only those micro-partitions holding the data required to satisfy the query results and excludes (or prunes) irrelevant micro-partitions.

Because the relevant metadata is known for each micro-partition, the contents can be compressed as the optimizer does not need to interrogate micro-partitions interactively to identify matching content. You also know each micro-partition contains up to 16 MB of data compressed using proprietary compression algorithms. Uncompressed, each micro-partition holds between 50 MB and 500 MB of data, and compression is optimized according to the column's data type.

Snowflake's internal, compressed, columnar format is not explained in detail. However, for the curious, I believe Snowflake utilizes the Partition Attributes Across (PAX) file format; you can find the whitepaper at `https://research.cs.wisc.edu/ multifacet/papers/vldb01_pax.pdf`.

Accessing Table Metadata

As you progress through this section, your investigation will identify programmatic approaches to determining the number of micro-partitions for a table. I do not offer a fully functional stored procedure-based approach but instead illustrate various methods to identify information of interest.

Using the Information Schema

Note that the information schema views are specific to an individual database. The next query illustrates the available information for tables:

```
SELECT *
FROM   tpc.information_schema.tables
WHERE  table_name LIKE '%_BASELINE%'
LIMIT  10;
```

There is no latency for information schema views, and dropped object information is not available.

The attribute `clustering_key` is NULL for tables declared without an explicit cluster key, discussed in detail within the next chapter.

Alternatively, you can use the SHOW command as this next SQL statement illustrates:

```
SHOW TABLES LIKE '%_BASELINE%';
```

Then convert the output to usable output by modifying the next SQL statement according to your needs:

```
SELECT "name",
       "rows",
       "automatic_clustering"
FROM   TABLE ( RESULT_SCAN ( last_query_id()));
```

Despite the latency inherent within the Account Usage Store views, you prefer to reference the Account Usage Store views as all account object information is available centrally. Alternatively, you would need to identify each database and access each information schema individually. Of course, your role would need to be entitled for each database, which may prove challenging in a multi-tenant or highly segregated Snowflake environment.

Using the Account Usage Store

Some table metadata is available from the Account Usage store as this next query illustrates:

```
SELECT  *
FROM    snowflake.account_usage.tables
WHERE   table_name LIKE '%_BASELINE%'
AND     deleted IS NULL
LIMIT   10;
```

Account Usage Store views experience data latency of between 45 minutes and 3 hours; therefore, you may not immediately see the expected results.

The previous query output contains useful information, and you will return to this content later; however, there is no mention of the number of micro-partitions.

Under what circumstances would it be useful to know the number of micro-partitions? And if you have a valid use case, how can you identify the required information?

One use case is replicating data between accounts where costs vary according to the number of micro-partitions transferred between primary and secondary accounts. While there are techniques explained later in this book to reduce the number of micro-partitions transferred, here you have a reason to know how many micro-partitions belong to each replicated object as each replica incurs storage cost.

With your use case defined and knowing each micro-partition contains compressed data, you can readily identify that using row counts as a proxy to derive micro-partition counts is not a valid approach.

All is not lost; there are other methods by which you can derive table micro-partition count:

- Using a query profile

- Using GET_QUERY_OPERATOR_STATS

- Using system$clustering_depth and system$clustering_information

You will now examine each in turn.

Query Profile

By issuing a simple query and accessing the query profile, you can readily identify the number of micro-partitions belonging to a table.

Let's first set the warehouse to X-Large and then create the example query.

```
USE WAREHOUSE IDENTIFIER ( $tpc_warehouse_xl );

SELECT *
FROM   lineitem_baseline;
```

Then click the query_id, which opens a new tab displaying the Statistics with a partition count of 9400 and, as an aside, spills to disk as indicated by both Local Disk I/O and Remote Disk I/O. Both metrics are shown in Figure 4-4.

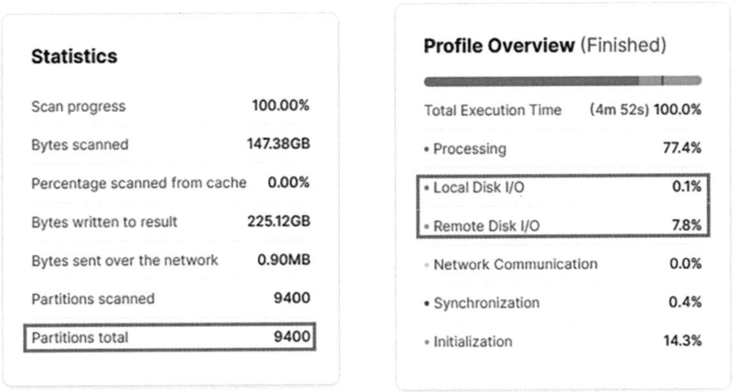

Statistics

Scan progress	100.00%
Bytes scanned	147.38GB
Percentage scanned from cache	0.00%
Bytes written to result	225.12GB
Bytes sent over the network	0.90MB
Partitions scanned	9400
Partitions total	9400

Profile Overview (Finished)

Total Execution Time	(4m 52s) 100.0%
• Processing	77.4%
• Local Disk I/O	0.1%
• Remote Disk I/O	7.8%
• Network Communication	0.0%
• Synchronization	0.4%
• Initialization	14.3%

Figure 4-4. *Query profile micro-partition count*

Then reset your warehouse to X-Small:

```
USE WAREHOUSE IDENTIFIER ( $tpc_warehouse_xs );
```

While the presented information is useful for an individual table query, the approach cannot be used for joins as the partitions total is the sum of all referenced tables. Furthermore, you cannot programmatically use the screen output to derive information from multiple tables; what you need is something more sophisticated.

GET_QUERY_OPERATOR_STATS

Utilizing the same query_id from the previous example query, let's examine the GET_ QUERY_OPERATOR_STATS output. Note that this feature is at Public Preview status at the time of writing.

```
SELECT *
FROM   TABLE ( get_query_operator_stats('01ae63a0-0000-8905-0000-000113
1e50ad'));
```

Within the returned data set, there are two attributes of interest. Within OPERATOR_ TYPE you are looking for the row with TableScan, and for the same row you use the OPERATOR_STATISTICS attribute to derive the partitions_total attribute.

```
{
  "io": {
    "bytes_scanned": 2048,
    "percentage_scanned_from_cache": 0,
    "scan_progress": 1
  },
  "output_rows": 5,
  "pruning": {
    "partitions_scanned": 1,
    "partitions_total": 1
  }
}
```

You might restate your GET_QUERY_OPERATOR_STATS SQL statement to extract only the partitions_total attribute as follows:

```
SELECT operator_statistics:pruning:partitions_total
FROM    TABLE ( get_query_operator_stats('01ae63a0-0000-8905-0000-000113
1e50ad'));
```

You will not see the partitions_total attribute if results are derived from the cache. The OPERATOR_TYPE will show QUERY_RESULT_REUSE.

This approach cannot be used for joins as the partitions total if the sum of all referenced tables and the OPERATOR_STATISTICS output differs. I will leave this for your further investigation.

You can find more information on GET_QUERY_OPERATOR_STATS at https://docs. snowflake.com/en/sql-reference/functions/get_query_operator_stats.

system$clustering_depth and system$clustering_information

The remaining option for determining micro-partition counts introduces a new system call that you will become very familiar with in the next chapter: system$clustering_depth.

I deliberately omit a full explanation of system$clustering_depth here and restrict usage to identifying the micro-partition count only.

You can assume your target table has not been clustered; I discuss cluster keys in the next chapter.

To prove my assumption, let's omit the column information.

```
SELECT system$clustering_depth ( 'LINEITEM_BASELINE' );
```

You should see this error message: "000005 (XX000): Invalid clustering keys or table LINEITEM_BASELINE is not clustered."

You can identify the number of micro-partitions within your target by adding a second parameter containing a single table attribute, as shown next:

```
SELECT system$clustering_depth ( 'LINEITEM_BASELINE', '(L_COMMENT)' );
```

The named attribute is not important so long as the attribute exists on the target table. Both table name and attribute name are case insensitive as I show later.

The query should return 9,398 micro-partitions.

You can also use `system$clustering_information` to derive the micro-partition count for an object:

```
SELECT system$clustering_information ( 'LINEITEM_BASELINE', '(L_
LINENUMBER)' );
```

The query should return a single row containing a JSON record. Look for an attribute named `total_partition_count`.

Alternatively, run this query to extract the `total_partition_count`:

```
SELECT parse_json(system$clustering_information ( 'lineitem_baseline', '(l_
shipdate)' )):total_partition_count;
```

In the previous examples, the table attribute must be enclosed by parenthesis, i.e., (L_COMMENT); otherwise, the query fails.

As you can see from both sample queries, this is a single-table approach to identifying micro-partitions.

You can find more information on `system$clustering_depth` and `system$clustering_information` at https://docs.snowflake.com/en/sql-reference/functions/system_clustering_depth and https://docs.snowflake.com/en/sql-reference/functions/system_clustering_information.

Time Sensitivity

Some of the SQL statements in this section are time sensitive. The ability to undrop objects and view how storage transitions from `active_GB` to `time_travel_GB` and then to `fail_safe_GB` are all dependent upon the following:

- The data retention period set at the database and/or object level

- Account Usage Store view latency

- Elapsed time between issuing commands

During the writing of this chapter, observing transitions between the differing layers of storage has not proven to be straightforward. I assume Snowflake does not run its processes to transition micro-partitions with high frequency and speculate its processes may execute at least once within the 90-minute view latency period. As a consequence, I did not experience a consistent up-to-the minute timeline for observing micro-partition transitions across storage layers; rather, the timeline was somewhat variable.

Your testing may be impacted if conducted over several days.

To illustrate time sensitivity, let's examine an idealized scenario and timeline, as represented in Figure 4-5.

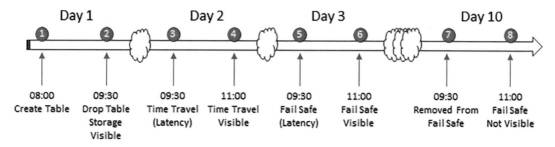

Figure 4-5. *Expected storage timeline with latency*

Here's a breakdown of Figure 4-5:

1. **Day 1:** At 08:00 you assume a table is created with the data retention period set to one day.

2. **Day 1:** At 09:30 view latency is complete, storage is visible, and then the table is dropped.

3. **Day 2:** At 09:30 the dropped table storage moves to Time Travel for one day, and the view has 90 minutes of latency.

4. **Day 2:** At 11:00 the view latency is complete, and Time Travel storage is visible.

5. **Day 3:** At 09:30 the dropped table storage moves to Fail-Safe for seven days, and the view has 90 minutes of latency.

6. **Day 3:** At 11:00 the view latency is complete, and Fail-Safe storage is visible.

7. **Day 10:** At 09:30 the dropped table storage exits Fail-Safe, and the view has 90 minutes of latency.

8. **Day 10:** At 11:00 the view latency is complete, and the Fail-Safe storage clears.

As exposed by the previous explanation, monitoring storage is not straightforward. Throughout the remainder of this section, you work through the expected storage timeline.

Data and Micro-partition Lifecycle

In this section I cover the performance time costs for INSERT, UPDATE, and DELETE. Recognize I have deliberately chosen to use an X-Small warehouse. I also expose the hidden storage costs of Time Travel and Fail-Safe before discussing storage implications for both cloning and replication.

Tune the design before implementing a single line of code.

Setting a Baseline

Using the view v_table_storage_metrics, let's establish the starting point by identifying your initial storage metrics for LINEITEM_BASELINE.

```
SELECT active_GB,
       time_travel_GB,
       failsafe_GB,
       retained_for_clone_GB
FROM   v_table_storage_metrics
WHERE  path_to_object = 'TPC.TPC_OWNER.LINEITEM_BASELINE';
```

Figure 4-6 shows the expected outcome where all values are set to zero except active_GB.

ACTIVE_GB	TIME_TRAVEL_GB	FAILSAFE_GB	RETAINED_FOR_CLONE_GB
147.377965450287	0.000000000000	0.000000000000	0.000000000000

Figure 4-6. *LINEITEM_BASELINE expected storage*

For the next suite of tests where you UPDATE, INSERT, and then DELETE data, I indicate the execution time differences for each operation, show the impact of deliberately using an X-Small warehouse, and finally expose how retained storage for Time Travel can affect your costs.

Most data warehouses are predicated upon periodic ingestion of data, and there are some use cases where the volume, velocity, and variety of change cause issues. As you may infer from the v_table_storage_metrics declaration, additional attributes provide deeper insight into object storage.

You must consider object data maintenance from several perspectives:

- Ingestion where you INSERT, UPDATE, and DELETE your data sets.

- Internal processing where you manipulate your staged data.

- Consumption where you SELECT, join, summarize, aggregate, and filter your data sets.

- The impact of object or database Time Travel setting.

- How Fail-Safe retains micro-partitions and consumes storage.

- Where cloned objects rely upon retained storage for deleted objects.

You will examine the impact of each perspective next.

Data Ingestion

Typically, you ingest data into a suite of staging table; let's ignore external tables for the purposes of this discussion.

Our data ingestion pattern should be optimized for single feeds or whole-schema ingestion. You should not expect to run reports or analytics workloads against data ingested into staging tables.

I must also address how DML operations affect micro-partition maintenance and this leads to tuning your ingestion design.

Depending upon volume, frequency of ingest, and subsequent DML operations to merge ingested data from your staging tables into your core schema, you may experience a high degree of micro-partition churn. Your staging tables will likely be Flush and Fill where each staging table is truncated and loaded. Pre-sorting your source data attributes into optimal columnar format before load can optimize onward processing though any benefits for smaller loads are likely to be minimal.

In some cases where data ingestion is a bottleneck, you have a few design options to consider.

- Adopt an insert-only model, i.e., Data Vault 2.0.

- Parallelize loads where discrete data isolation boundaries can be enforced.

- Reduce batch size and increase frequency noting the likely increase in micro-partition churn.

- Increase warehouse size.

- When sourcing data from external CSP storage, implement file caching or faster storage devices.

We will return to parallelization in a later chapter as the subject is worthy of wider consideration.

In the next section, you will investigate storage costs, but for now, you might consider optimizing your data ingestion costs by doing the following:

- Setting Time Travel to 0 for your staging tables where persistence is not required; note Fail-Safe is retained at 7 days.

- Use transient tables for your staging tables with Time Travel set to 0 as transient tables do not utilize Fail-Safe.

Both of these approaches imply staged data can be reloaded from source. Where external stages are used, ensure the CSP storage is set to retain files according to the requirements.

Data Processing

With your data loaded into staging tables and all feed dependencies resolved, core application processing occurs.

In this section, you will investigate "how" to identify both performance timings and storage costs.

Typical data processing operations are to merge staged data into target tables using MERGE, INSERT, UPDATE, and DELETE operators.

For a 2.5-billion row unclustered table, I found INSERT and DELETE operations to perform best. UPDATE operations were problematic and took much longer to complete.

Let's investigate these scenarios further. You start with timing an UPDATE operation using an X-Small warehouse, which I timed at over 26 minutes to complete; you may want to experiment with a larger warehouse sizing. The UPDATE will affect 3,000,013,782 records.

```
USE WAREHOUSE IDENTIFIER ( $tpc_warehouse_xs );

UPDATE tpc.tpc_owner.lineitem_baseline
SET    l_linestatus = 'X'
WHERE  l_linestatus = 'F';
```

Let's see the effect on storage by repeating the earlier query noting that you may experience latency.

```
SELECT active_GB,
       time_travel_GB,
       failsafe_GB,
       retained_for_clone_GB
FROM   v_table_storage_metrics
WHERE  path_to_object = 'TPC.TPC_OWNER.LINEITEM_BASELINE';
```

Figure 4-7 shows two rows.

- The first row is the original expected data set size.

- The second row is after the UPDATE statement. Note that TIME_TRAVEL_GB reflects the storage retained for the UPDATED rows.

ACTIVE_GB	TIME_TRAVEL_GB	FAILSAFE_GB	RETAINED_FOR_CLONE_GB
147.377965450287	0.000000000000	0.000000000000	0.000000000000
147.624543666840	93.925645351410	0.000000000000	0.000000000000

Figure 4-7. *UPDATE storage consumption*

You might also want to compare the number of micro-partitions for lineitem_
baseline.

```
SELECT parse_json(system$clustering_information ( 'lineitem_baseline',
'(l_shipdate)' )):total_partition_count;
```

The previous query should return 9405; the initial micro-partition count was 9398. From your investigation, three facts emerge.

- The amount of storage required to hold the same amount of data has increased.

- The micro-partition count has increased from 9398 to 9405.

- Time Travel storage has been created; this is expected behavior.

You should expect a single character update to use the *same* amount of storage as the original data set, but the evidence proves there has been an increase in storage. you should also expect the number of micro-partitions for both before and after your UPDATE. Why do your figures not match?

The answer may lie with how the internal cluster key has been declared, and you know from the earlier discussion it is not possible to determine the internal clustering key attribute order. I discuss cluster keys in the next chapter.

Let's investigate these scenarios further; I started with timing an UPDATE operation using an X-Small warehouse, which I timed at more than 18 minutes to complete. My INSERT created 3,000,013,782 records and took around 8 minutes less to complete than an UPDATE.

```
INSERT INTO tpc.tpc_owner.lineitem_baseline
SELECT *
FROM    snowflake_sample_data.tpch_sf1000.lineitem
WHERE   l_linestatus = 'F';
```

Then check the effect on storage by repeating the earlier query; note that you may experience latency.

```
SELECT active_GB,
       time_travel_GB,
       failsafe_GB,
       retained_for_clone_GB
```

```
FROM    v_table_storage_metrics
WHERE   path_to_object = 'TPC.TPC_OWNER.LINEITEM_BASELINE';
```

Figure 4-8 shows three rows.

- The first row is the original expected data set size.

- The second row is after the UPDATE statement.

- The third row is after the INSERT statement; note that time_travel_
  GB reflects the storage retained for the UPDATE rows.

ACTIVE_GB	TIME_TRAVEL_GB	FAILSAFE_GB	RETAINED_FOR_CLONE_GB
147.377965450287	0.000000000000	0.000000000000	0.000000000000
147.624543666840	93.925645351410	0.000000000000	0.000000000000
221.281464099884	93.925645351410	0.000000000000	0.000000000000

Figure 4-8. *INSERT storage consumption*

You know that new micro-partitions will have been added due to the INSERT, the new data volume is reflected in the active_GB column for the third record. As you are inserting new records, no Time Travel storage is created; this is expected behavior.

You expect the number of micro-partitions to be about 50% more than 9405:

```
SELECT parse_json(system$clustering_information ( 'lineitem_baseline', '(l_
shipdate)' )):total_partition_count;
```

The returned micro-partition count is 14,074 indicating 4,669 new micro-partitions were created in line with expectations.

The last SQL statement is a DELETE operation using an X-Small warehouse, which I timed at more than six minutes to complete. The DELETE removed 3,000,013,782 records and took around 20 minutes less to complete than the UPDATE and 12 minutes less than the INSERT.

```
DELETE FROM tpc.tpc_owner.lineitem_baseline
WHERE   l_linestatus = 'X';
```

Then check the effect on storage by repeating the earlier query; you may experience latency:

```
SELECT active_GB,
       time_travel_GB,
       failsafe_GB,
       retained_for_clone_GB
FROM   v_table_storage_metrics
WHERE  path_to_object = 'TPC.TPC_OWNER.LINEITEM_BASELINE';
```

Figure 4-9 shows four rows.

- The first row is the original expected data set size.

- The second row is after the UPDATE statement.

- The third row is after the INSERT statement.

- The fourth row shows the effect of the DELETE statement; note that time_travel_GB increase due to the storage retained for the DELETED rows.

ACTIVE_GB	TIME_TRAVEL_GB	FAILSAFE_GB	RETAINED_FOR_CLONE_GB
147.377965450287	0.000000000000	0.000000000000	0.000000000000
147.624543666840	93.925645351410	0.000000000000	0.000000000000
221.281464099884	93.925645351410	0.000000000000	0.000000000000
147.125027656555	186.928283691406	0.000000000000	0.000000000000

Figure 4-9. *DELETE storage consumption*

You know micro-partitions will have been replaced due to the DELETE; the new data volume is reflected in the active_GB column for the fourth record. As you are deleting records, Time Travel storage is retained; this is expected behavior.

You can expect the number of micro-partitions to be about 9,405.

```
SELECT parse_json(system$clustering_information ( 'lineitem_baseline',
'(l_shipdate)' )):total_partition_count;
```

The returned micro-partition count is 9,362 indicating 4,712 micro-partitions were removed in line with your expectations.

The simple walk-through of the life cycle of data supports the earlier assertion of both `INSERT` and `DELETE` operations being faster than `UPDATE` operations. Your mileage may vary according to the volume, velocity, and variety of change experienced within your application; note the retained data indicated by `time_travel_GB` can far exceed initial expectations.

Unlike data ingestion and data consumption, for data processing, you have limited options when considering how to reduce storage costs.

- Set Time Travel to the minimum required period for each object.

- Use temporary tables and transient tables where possible.

- Use clones noting that new micro-partitions are created when data changes within the cloned object.

Snowflake recommends the use of temporary tables to hold intermediate result sets to reduce query complexity.

Data Consumption

In the previous section, I discussed data processing where you ingest data from your staging tables and ingest into your core application components. You also began to see the impact of data retention in support of Time Travel.

The same considerations apply to outbound data consumption where you will observe storage is retained due to your data retention period, particularly when moving data from third normal form to Data Vault 2.0 and then into star schemas with each data model retaining a copy of all data. The important point to note is the physical cost of storing multiple copies of data within your various models along with the hidden cost of storage to support both Time Travel setting and Fail-Safe retention period.

Data consumption can lead to increased storage costs where you may need to denormalize data. The same storage considerations apply when you create these objects as you consume additional storage to support faster and more user-friendly data access paths. You do not get anything for free, and system implementation is usually a trade-off. There is always a price to pay either in terms of performance or storage.

Where data sets are periodically rebuilt and history is not required, you might consider optimizing your data consumption costs by doing the following:

- Set Time Travel to 0 for periodically rebuilt tables where persistence is not required; note that Fail-Safe is retained for seven days.

- Use transient tables for periodically rebuilt tables with Time Travel set to 0 as transient tables do not utilize Fail-Safe.

Both of these approaches imply periodically rebuilt tables can be rebuilt from source within acceptable timeframes and with minimal business impact.

Time Travel

In previous sections many references have been made to Time Travel, and you have seen the impact of UPDATE, INSERT, and DELETE operations on data retention too.

Many Snowflake applications set Time Travel at the database level to 90 days ensuring that all objects created within the database inherit the default setting. With your new understanding of the storage implications for high Time Travel retention settings, you must adopt a more nuanced approach.

Not all applications are equal; your requirements will differ accordingly, and the key takeaway from the next suite of suggestions is to balance your efforts. Storage is relatively cheap these days.

To assist tuning your storage design, I list some options for your consideration; again, remember not to "boil the ocean." Focus on the cheapest and quickest options to return the maximum benefit for the minimum amount of expended effort.

- Where ingested data can easily be reloaded, choose either temporary or transient tables.

- Where processed data is subject to high-frequency, low-volume DML activity, set Time Travel as low as acceptable.

- Build intermediate data sets into temporary tables before loading into core tables.

- Parallelize high-frequency, low-volume data loads to reduce micro-partition churn.

- Adopt an insert-only design pattern such as Data Vault 2.0.

- Where consumed data is periodically re-created, choose transient tables.

- For large tables, implement optimal cluster keys to match the most common data access paths; see the next chapter for details.

Let's first identify the Time Travel settings for your TPC database:

```
SELECT  retention_time
FROM    snowflake.account_usage.databases
WHERE   database_name = 'TPC'
AND     deleted IS NULL;
```

Our query should return 90 indicating the Time Travel retention period is 90 days for the TPC database.

Now create a table called `lineitem_baseline_tt_test` for immediate DROP; the Data Retention Period `data_retention_time_in_days` is set to 1.

```
USE WAREHOUSE IDENTIFIER ( $tpc_warehouse_xl );
```

The next statement `CREATE TABLE AS SELECT` (CTAS) will take a few minutes to run.

```
CREATE OR REPLACE TABLE tpc.tpc_owner.lineitem_baseline_tt_test
data_retention_time_in_days = 1
AS
SELECT *
FROM    tpc.tpc_owner.lineitem_baseline;
```

Immediately drop your new table, `lineitem_baseline_tt_test`.

```
DROP TABLE tpc.tpc_owner.lineitem_baseline_tt_test;
```

Then reset your warehouse to X-SMALL.

```
USE WAREHOUSE IDENTIFIER ( $tpc_warehouse_xs );
```

Then check the effect on storage by executing the following query; you may experience latency:

```
SELECT active_GB,
       time_travel_GB,
       failsafe_GB,
```

```
        retained_for_clone_GB
FROM    v_table_storage_metrics
WHERE   path_to_object
        = 'TPC.TPC_OWNER.LINEITEM_BASELINE_TT_TEST';
```

Figure 4-10 shows the retained but inaccessible storage for your created and then immediately dropped table.

ACTIVE_GB	...	TIME_TRAVEL_GB	FAILSAFE_GB	RETAINED_FOR_CLONE_GB
147.945022583008		0.000000000000	0.000000000000	0.000000000000

Figure 4-10. *Dropped table storage consumption*

While storage for the dropped table is retained as `active_GB`, you cannot access the table. You can prove the storage is inaccessible by attempting to `SELECT` from the dropped table, which will result in failure, an exercise I leave for your further investigation.

You can find more information on Time Travel at `https://docs.snowflake.com/user-guide/data-time-travel`.

Data retained for Time Travel will later transition into Fail-Safe, which I discuss shortly.

Recovered Objects

You can recover the most recent version of a table using the `UNDROP` command for objects dropped within the Data Retention Period, as this example shows:

```
UNDROP TABLE tpc.tpc_owner.lineitem_baseline_tt_test;
```

Prove you can access the data from your recovered object:

```
SELECT *
FROM    tpc.tpc_owner.lineitem_baseline_tt_test
LIMIT 10;
```

To continue your investigation into Fail-Safe, you now `DROP` your test table again:

```
DROP TABLE tpc.tpc_owner.lineitem_baseline_tt_test;
```

Fail-Safe

Fail-Safe is an immutable seven-day period where micro-partitions *on a best-effort basis* are retained for recovery with the assistance of Snowflake Support. Fail-Safe is a last-resort; data is not accessible by any users.

Having dropped the test table in the previous section, you must wait until the dropped micro-partitions transition through Time Travel into Fail-Safe. Note the Data Retention Period for `lineitem_baseline_tt_test` was declared to be one day. You can repeat the earlier query to check the storage:

```
SELECT active_GB,
       time_travel_GB,
       failsafe_GB,
       retained_for_clone_GB
FROM   v_table_storage_metrics
WHERE  path_to_object
          = 'TPC.TPC_OWNER.LINEITEM_BASELINE_TT_TEST';
```

Figure 4-11 shows your micro-partitions have transitioned to Fail-Safe.

The first row is your original image from Figure 4-10.

- The second row is the new Fail-Safe data.

ACTIVE_GB	⋯	TIME_TRAVEL_GB	FAILSAFE_GB	RETAINED_FOR_CLONE_GB
147.945022583008		0.000000000000	0.000000000000	0.000000000000
0.000000000000		0.000000000000	147.945022583008	0.000000000000

Figure 4-11. *Fail-safe storage consumption*

You expect `failsafe_GB` to be the same as the original `active_GB` as Figure 4-11 demonstrates.

To recover data retained within Fail-Safe or raise a support ticket, see `https://community.snowflake.com/s/article/How-To-Submit-a-Support-Case-in-Snowflake-Lodge`.

You can find more information on Fail-Safe at `https://docs.snowflake.com/en/user-guide/data-failsafe` and at `https://docs.snowflake.com/en/user-guide/data-cdp-storage-costs`.

Cloned Objects

The cost of maintaining cloned objects is rarely discussed yet can contribute significant storage costs. At the point of initial cloning, the original "parent" and cloned "child" share the same micro-partitions. Assuming both parent and child are subject to different DML actions using disparate data, the parent and child tables will diverge in content. But what happens to the micro-partitions?

Micro-partitions for the parent will be superseded as expected with the full lineage preserved according both Time Travel setting and Fail-Safe period.

Micro-partitions for the child untouched by DML activity for either parent or child remain referenced back to the parent.

Figure 4-12 on the left shows both parent and child referencing the same micro-partitions after cloning. On the right is the effect of DML activity to both the parent and child showing how:

- Micro-partitions are created where contents diverge.

- Micro-partitions are moved to Time Travel where contents are superseded.

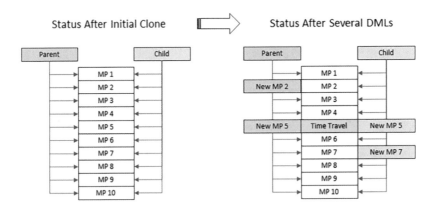

Figure 4-12. *Cloned object storage consumption*

Let's investigate cloned object storage consumption using a practical example. You will use `partsupp_baseline` and first check the allocated storage:

```
SELECT path_to_object,
       active_GB,
       time_travel_GB,
```

```
          failsafe_GB,
          retained_for_clone_GB
FROM      v_table_storage_metrics
WHERE     path_to_object = 'TPC.TPC_OWNER.PARTSUPP_BASELINE';
```

Figure 4-13 shows the expected result. Only active_GB storage is allocated.

PATH_TO_OBJECT	ACTIVE_GB	TIME_TRAVEL_GB	FAILSAFE_GB	RETAINED_FOR_CLONE_GB
TPC.TPC_OWNER.PARTSUPP_BASELINE	27.552062988281	0.000000000000	0.000000000000	0.000000000000

Figure 4-13. *partsupp_baseline active storage*

You now clone partsupp_baseline to partsupp_baseline_clone.

```
CREATE TABLE tpc.tpc_owner.partsupp_baseline_clone
CLONE   tpc.tpc_owner.partsupp_baseline;
```

Now re-check consumed storage for both the parent and child tables.

```
SELECT path_to_object,
       active_GB,
       time_travel_GB,
       failsafe_GB,
       retained_for_clone_GB
FROM   v_table_storage_metrics
WHERE  path_to_object IN
         ( 'TPC.TPC_OWNER.PARTSUPP_BASELINE',
           'TPC.TPC_OWNER.PARTSUPP_BASELINE_CLONE' );
```

The results should be identical to Figure 4-13 as shown indicating no additional storage has been allocated for partsupp_baseline_clone.

Automatic clustering is suspended for cloned tables.

In preparation for examining how updates to cloned tables affect storage, you pick a random ps_suppkey and count how many rows will be affected.

```
SELECT count(1)
FROM    partsupp_baseline_clone
WHERE   ps_suppkey = 1305848;
```

You should see 80 rows returned.

You cannot know how many micro-partitions will be affected at this point, but for reference, let's identify the current micro-partition count for both parent and child tables. You expect the returned counts to be identical as each object references the same micro-partitions.

```
SELECT parse_json(system$clustering_information ( 'partsupp_baseline',
'(ps_suppkey)' )):total_partition_count;
```

```
SELECT parse_json(system$clustering_information ( 'partsupp_baseline_
clone', '(ps_suppkey)' )):total_partition_count;
```

Both return 1,679 micro-partitions.

Now update your clone table called partsupp_baseline_clone using your random ps_suppkey value.

```
UPDATE tpc.tpc_owner.partsupp_baseline_clone
SET     ps_comment = 'Clone Test'
WHERE   ps_suppkey = 1305848;
```

You should see 80 rows updated.

And check whether storage has been affected by the UPDATE; you may experience latency:

```
SELECT path_to_object,
       active_GB,
       time_travel_GB,
       failsafe_GB,
       retained_for_clone_GB
FROM   v_table_storage_metrics
WHERE  path_to_object IN
          ( 'TPC.TPC_OWNER.PARTSUPP_BASELINE',
            'TPC.TPC_OWNER.PARTSUPP_BASELINE_CLONE' );
```

Figure 4-14 shows the expected result. Only `active_GB` storage is allocated for active or current micro-partitions:

PATH_TO_OBJECT	ACTIVE_GB	TIME_TRAVEL_GB	FAILSAFE_GB	RETAINED_FOR_CLONE_GB
TPC.TPC_OWNER.PARTSUPP_BASELINE_CLONE	0.033934116364	0.000000000000	0.000000000000	0.000000000000
TPC.TPC_OWNER.PARTSUPP_BASELINE	27.552062988281	0.000000000000	0.000000000000	0.000000000000

Figure 4-14. *partsupp_baseline and partsupp_baseline_clone Active Storage*

As you see, `active_GB` for `partsupp_baseline_clone` demonstrates new micro-partitions have been allocated for your updated data.

You might also re-check the number of micro-partitions for both the parent and child; this may not be either informative or conclusive. You must remember it is not possible to determine the internal clustering key attribute order. As you have only updated the child table, the parent table micro-partition count will be constant. You expect 1,679 micro-partitions to be returned by the next query:

```
SELECT parse_json(system$clustering_information ( 'partsupp_baseline',
'(ps_suppkey)' )):total_partition_count;
```

And for `partsupp_baseline_clone`, you might see the same number of micro-partitions, more, or fewer depending upon whether clustering is affected by the UPDATE. In my environment, the next SQL statement returned the same 1,697 micro-partition count as before:

```
SELECT parse_json(system$clustering_information ( 'partsupp_baseline_
clone', '(ps_suppkey)' )):total_partition_count;
```

Our next objective is to illustrate how storage is retained for cloned objects when the parent is dropped. First, confirm `retention_time` is set to one day.

```
SELECT table_name,
       retention_time
FROM   tpc.information_schema.tables
WHERE  table_name LIKE '%LINEITEM_BASELINE%';
```

You expect both `partsupp_baseline` and `partsupp_baseline_clone` are set to one day; if not, issue the next SQL statement.

```
ALTER TABLE tpc.tpc_owner.partsupp_baseline
SET          data_retention_time_in_days = 1;
```

Having confirmed the data retention period, let's drop the parent table.

```
DROP TABLE tpc.tpc_owner.partsupp_baseline;
```

And check whether storage has been affected by the earlier UPDATE; you may experience latency.

```
SELECT path_to_object,
       active_GB,
       time_travel_GB,
       failsafe_GB,
       retained_for_clone_GB
FROM   v_table_storage_metrics
WHERE  path_to_object IN
          ( 'TPC.TPC_OWNER.PARTSUPP_BASELINE',
            'TPC.TPC_OWNER.PARTSUPP_BASELINE_CLONE' );
```

Figure 4-15 shows the expected result where active_GB storage is allocated for active or current micro-partitions.

PATH_TO_OBJECT	ACTIVE_GB	TIME_TRAVEL_GB	FAILSAFE_GB	RETAINED_FOR_CLONE_GB
TPC.TPC_OWNER.PARTSUPP_BASELINE_CLONE	0.033934116364	0.000000000000	0.000000000000	0.000000000000
TPC.TPC_OWNER.PARTSUPP_BASELINE	27.552062988281	0.000000000000	0.000000000000	0.000000000000

Figure 4-15. partsupp_baseline and partsupp_baseline_clone Active Storage

You must wait at least one day before storage migrates to time_travel_GB after which time you expect to see the following:

- The active_GB value reduces as any unreferenced micro-partitions for the dropped parent table transition to time_travel_GB.

- time_travel_GB increases reflecting the unreferenced dropped parent table micro-partitions.

- When Time Travel data retention period for the dropped parent table expires, failsafe_GB to increase and time_travel_GB to decrease.

During your testing, after three days, you found the storage *did not* migrate to time_travel_GB. You suspect the default cluster key on the "parent" table did not cause micro-partitions to be de-referenced.

There is an important side effect. While you would expect the parent table retention period to age out old micro-partitions, you find the retained micro-partitions allow UNDROP operations, leading to the recovery of the `partsupp_baseline` table.

```
UNDROP TABLE tpc.tpc_owner.partsupp_baseline;
```

While I would not choose to rely upon unexpected micro-partition retention to UNDROP objects, I suggest this action may be possible and should not be discounted. Simply put, you cannot determine whether UNDROP will work until you try, and an attempt to UNDROP has a temporal component.

Data Sharing and Replication

Data sharing within the same CSP and region is implemented by sharing micro-partitions with consumers via Secure Direct Data Share, Private Listings, or Snowflake Marketplace. Consumers ingesting shared data reference current micro-partitions only; they see data producer transactions in real time at zero cost. Consumers cannot see any historical transactions, nor can they access Time Travel or Fail-Safe for the producer account.

Replication is implemented by shipping changed micro-partitions to consuming accounts using either database replication or account replication. Ingesting replicated micro-partitions requires replication to be configured, which is a timed refresh event and therefore not real time.

A full investigation of data sharing and replication is beyond the scope of this book and is worthy of a significant chapter on its own, which I delivered in my previous book, *Maturing the Snowflake Data Cloud.*

You can find more information at `https://docs.snowflake.com/en/guides-overview-sharing`.

Micro-partitions End to End

Throughout this chapter you have worked through how micro-partitions are both expected and observed to transition from active through Time Travel and Fail-Safe and then removal while also considering cloning.

You also investigated how the data retention period and Account Usage Store latency affects observability of transitions across each state. You also learned there are several unknown factors relating to the frequency at which both the Snowflake internal processes run and process interactions for information collation affect latency.

Micro-partitions are a difficult subject to address.

Based on just Snowflake-supplied information, I have found this chapter difficult to write, so here I present the best interpretation of the available evidence.

In general, you see the Snowflake state transition holds true, though observability proves difficult, if not impossible, to accurately define in time. Taking all factors into consideration, the closest analog is to say that observability is *eventually consistent* with expectations, but I cannot say exactly *when* consistency occurs.

In an ideal situation, you would observe the following behavior using mocked-up data.

First, create a table with `data_retention_time_in_days = 1`.

Examine storage using the view `v_table_storage_metrics`. Figure 4-16 shows the expected result where `active_GB` storage is allocated for the sample table.

ACTIVE_GB	TIME_TRAVEL_GB	FAILSAFE_GB	RETAINED_FOR_CLONE_GB
6.694985389709	0.000000000000	0.000000000000	0.000000000000

Figure 4-16. *Sample table active storage at creation*

After creation you update your table contents.

When latency has expired, you re-examine the storage. Figure 4-17 shows the expected result where the `active_GB` value has changed.

ACTIVE_GB	TIME_TRAVEL_GB	FAILSAFE_GB	RETAINED_FOR_CLONE_GB
6.695380687714	0.000000000000	0.000000000000	0.000000000000

Figure 4-17. *Sample table active storage after update*

When micro-partitions have transitioned into Time Travel and latency has elapsed, Figure 4-18 shows the expected result where the `time_travel_GB` value has changed.

ACTIVE_GB	TIME_TRAVEL_GB	FAILSAFE_GB	RETAINED_FOR_CLONE_GB
6.695380687714	0.033465385437	0.000000000000	0.000000000000

Figure 4-18. *Sample table time travel storage*

After both `data_retention_time_in_days` and latency have elapsed, Figure 4-19 shows the expected result where the `failsafe_GB` value has changed.

ACTIVE_GB	TIME_TRAVEL_GB	FAILSAFE_GB	RETAINED_FOR_CLONE_GB
6.695380687714	0.000000000000	0.033465385437	0.000000000000

Figure 4-19. *Sample table fail-safe storage*

After seven days, your micro-partitions are removed. Figure 4-20 shows the expected result where `failsafe_GB` value has changed back to zero.

ACTIVE_GB	TIME_TRAVEL_GB	FAILSAFE_GB	RETAINED_FOR_CLONE_GB
6.695380687714	0.000000000000	0.000000000000	0.000000000000

Figure 4-20. *Sample table micro-partition removal*

A similar sequence can be derived for cloned tables with updates.

Micro-partition Pitfalls

With Snowflake, the ability to clone and recover objects "at will" brings unforeseen challenges when managing your accounts.

- Developers and operations support staff forget to clean up temporary objects created during production releases and maintenance activities.

- Stages also consume storage.

    - Internal stages should be monitored for use and periodically removed where possible.

    - External stages consume CSP storage and likewise require periodic cleanup.

141

- Deleting an external stage does not remove files contained within the external stage.

- Incorrectly setting object data retention periods leads to excessive storage retention.

- Using permanent tables where transient or temporary tables are more cost effective.

- Database and object explosion where new environments are cloned for regression testing but never deleted.

- Cloning and Time Travel:

  - These make object retention too easy, leading to bad practices.

  - Reduced storage requirements when judiciously used for creating development and test environments.

Where the cost is zero, the demand is infinite. My observation is that controlling costs are typically focused on credit consumption and not on managing storage costs. Universally CSP storage is cheap, roughly $23/TB at the time of writing, though this figure is CSP and region specific. For small data footprints, the costs are almost insignificant, but at petabyte scale, the costs quickly escalate.

As your Snowflake usage increases and environment matures, I suggest the following:

- Implementing central storage monitoring

- Adopting guidelines for "acceptable use" of cloning and Time Travel

- Periodically reviewing environments to mitigate against increasing storage

- In multi-tenant environments, cross-charging each tenant for their storage in addition to their runtime consumption

Summary

In this chapter, I covered micro-partitions and different ways to identify the number of micro-partitions belonging to an object.

I then explained how time affects micro-partition observability along with a discourse on the idealized micro-partition life cycle.

Stepping through the traditional segments of an application life cycle illustrated the impact of incorrectly setting data retention periods. I then provided justification for using transient tables for specific components within your applications.

Our investigation into Time Travel, Fail-Safe, and cloning demonstrated hidden storage costs incurred by micro-partition retention. I then identified some challenges with uncontrolled cloning and made recommendations to mitigate such actions.

With a firm grasp of micro-partitions, appropriate object creation, use, and maintenance, I will next discuss cluster keys.

CHAPTER 5

Cluster Keys

Cluster keys go hand in hand with micro-partitions. It may not be evident that all object micro-partitions have a cluster key, but they do. Identifying the constituent attributes of an undeclared cluster key will prove difficult if not impossible to determine, but rest assured every micro-partition has a cluster key.

If you read the previous chapter, you have encountered some of the tooling used to identify micro-partitions counts and, as a side effect, will already have begun to gather some information about cluster keys.

In this chapter, you will do the following:

- Learn about the available tooling to extract cluster key information

- Learn how to determine optimal cluster key columns and ordering

- Investigate clustering to identify "good" and "bad" patterns

- Investigate materialized views and dynamic tables

Let's start this chapter by asking "What is a cluster key?" within the context of micro-partitions holding data in a hybrid columnar compressed format. Having established what cluster keys are and how they are identified, I will then discuss the implications for data storage and how the pruner implements micro-partition pruning.

If you read my previous book, *Building the Snowflake Data Cloud*, some of the content here will be familiar. While the core themes remain the same, we will dive deeper and investigate more widely than before.

Cluster keys have two key objectives: to enable efficient micro-partition pruning and to speed up data retrieval. You may also experience compression benefits due to repeating values, as you will see while testing later within this chapter. But there is a catch: your query predicates (the WHERE clauses) must match the cluster key attribute ordering for maximum efficiency, as covered in this chapter.

A. Carruthers, *Tuning the Snowflake Data Cloud*, https://doi.org/10.1007/979-8-8688-0379-6_5

Restating the foundational principle of micro-partition immutability provides insights into some performance issues; note that the following list relates to micro-partitions and must be considered in addition to the root causes already identified. In no fixed order and not exhaustive, these factors could contribute to poor performance:

- Missing cluster keys for query predicates

- Mismatched cluster keys for query predicates

- High-frequency, low-volume updates

- Parallel or concurrent DML against the same object

- Data skewed against original expectations

- Mismatched warehouses for the requested operation resulting in spills to disk or out-of-memory (OOM) errors

Having identified a few potential root causes of performance issues, I also acknowledge cluster keys are not the only performance tuning option available.

Finally, I will show the optimal use of the Snowflake automatic clustering feature in real-world scenarios along with its cost, time, storage, and performance implications.

Foundational Information

In this section, I describe the basic concepts relating to cluster keys; I start with a refresher on cardinality before moving on to micro-partition counts.

I will explain two related principles fundamental to understanding micro-partition pruning: clustering width and clustering depth.

Using both clustering width and clustering depth, I will then cover cluster metrics using an example date.

Cardinality

Within relational database management systems (RDBMSs), *cardinality* relates to the number of repeat elements within a data set.

If an attribute contains identical values, the cardinality must be 1. Conversely, if an attribute contains two or more different values, the cardinality will be greater than 1.

You can assume your Snowflake session context has been set as previously described.

```
SELECT  r_regionkey,
        r_name
FROM    region_baseline
ORDER BY 1 ASC;
```

Figure 5-1 shows the result set has unique entries for both the r_regionkey and r_name attributes; therefore, the cardinality is 5 for both r_regionkey and r_name.

R_REGIONKEY	R_NAME
0	AFRICA
1	AMERICA
2	ASIA
3	EUROPE
4	MIDDLE EAST

Figure 5-1. *Region cardinality*

Ideal attributes for cluster keys have a low number of values, with the corresponding table containing a high record count.

For the more mathematically inclined, Wikipedia provides a far more detailed explanation at https://en.wikipedia.org/wiki/Cardinality.

Micro-partition Counts

Micro-partition counts are an important metric to capture to identify whether pruning has occurred when executing SQL statements.

I am assuming your Snowflake session context has been set as previously described.

As a reminder of how to derive micro-partition counts, in the following SQL statement the table attribute must be enclosed by parentheses: (L_COMMENT). Otherwise, the query fails for a nonexplicitly clustered table, as this example shows:

```
SELECT system$clustering_depth ( 'LINEITEM_BASELINE' );
```

You should see this error message: "000005 (XX000): Invalid clustering keys or table LINEITEM_BASELINE is not clustered."

You can identify the number of micro-partitions within your target by adding a second parameter containing a single table attribute. Note that not all attribute names are equally effective. Avoid DATE attributes.

```
SELECT system$clustering_depth ( 'LINEITEM_BASELINE', '(L_COMMENT)' );
```

Figure 5-2 shows the expected result.

SYSTEM$CLUSTERING_DEPTH ('LINEITEM_BASELINE', '(L_COMMENT)')

9,406

Figure 5-2. *Cluster depth micro-partition count*

You can also use `system$clustering_information` to derive the micro-partition count for an object. Note that I will cover `system$clustering_information` more deeply later.

```
SELECT parse_json(system$clustering_information ( 'lineitem_baseline',
'(l_shipdate)' )):total_partition_count;
```

Figure 5-3 shows the expected result.

PARSE_JSON(SYSTEM$CLUSTERING_INFORMATION ('LINEITEM_BASELINE', '(L_SHIPDATE)')):TOTAL_PARTITION_COUNT

9406

Figure 5-3. *Cluster information micro-partition count*

I will assume you are familiar with the previous query outputs.

I will cover how to use these same functions to identify cluster keys later in the chapter.

Clustering Ratio

Clustering ratio is a deprecated measure expected to be removed soon. While the term may still appear in the Snowflake documentation, there is no clear definition of the formula used to calculate this measure. Therefore, I suggest that clustering ratio is no longer relevant, but for the curious, you can find more information at `https://docs.snowflake.com/en/sql-reference/functions/system_clustering_ratio`.

Cluster Width

Cluster width is the number of micro-partitions for a given object where the micro-partition content overlaps with other micro-partition content. Cluster width is a *horizontal* measure used to indicate pruning strategy effectiveness. A high value for the cluster width indicates the micro-partition contents overlap with several other micro-partition contents.

You want to reduce the cluster width value to 1, which indicates that the micro-partition content does not overlap. This represents the optimal access path for micro-partition pruning.

Cluster Depth

Cluster depth is the number of micro-partitions where any given attribute value overlaps with other micro-partitions. Cluster depth is a *vertical* measure derived for each attribute with the maximum derived number used for cluster depth. A high value for the cluster depth indicates that a given attribute value exists in many other micro-partitions.

You aim to reduce cluster depth value to 1, indicating that the attribute values do not overlap. This represents the optimal access path for micro-partition pruning. The average depth of a populated table (i.e., a table containing data) is always 1 or more.

You can find more information on cluster depth at `https://docs.snowflake.com/en/user-guide/tables-clustering-micropartitions#label-clustering-depth`.

Illustrating Cluster Width and Cluster Depth

Your objective, for any given query predicates, is for the optimizer to access a single (or very few) micro-partitions. In practice, the lower the number of overlapping micro-partitions, the better, as this indicates queries will be more selective. In other words, you should aim for each micro-partition to contain a range of minimum and maximum values for the cluster key.

To illustrate both cluster width and cluster depth, Figure 5-4 shows a table comprising four micro-partitions (MP1 to MP4) containing values A to Z. Assume the A to Z values are spread unevenly across all four micro-partitions with overlaps.

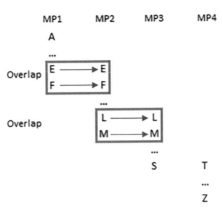

Figure 5-4. *Overlapping source data*

Figure 5-5 shows both cluster width and cluster depth for a single table with four micro-partitions where:

- The cluster width shows three micro-partitions containing E/F and L/M overlapping. The cluster width is the sum of *all* overlapping micro-partitions.

- The cluster depth shows two pairs of two attributes (E/F and L/M) overlapping. The cluster depth is the highest value of E/F or L/M attribute overlap in micro-partitions; in this example, both values are 2.

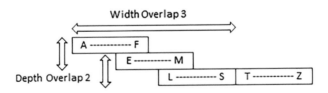

Figure 5-5. *Illustrating cluster width and cluster depth*

In an ideal situation, there would be no overlaps for cluster width or cluster depth. In real-world situations, you will often find overlaps, though. Your objective is to reduce overlaps to as close to 1 as possible in both dimensions.

Cluster Key Basics

Before we progress with this chapter, let's answer the question, "What is a cluster key?"

What Is a Cluster Key?

According to the Snowflake documentation, "A clustering key is a subset of columns in a table (or expressions on a table) that are explicitly designated to co-locate the data in the table in the same micro-partitions."

The absence of a cluster key is not necessarily a "bad" thing. Some tables are too small to benefit from adding a cluster key. Snowflake recommends adding cluster keys for tables of 1TB or larger, but this is not a hard-and-fast rule. As with every attempted performance tuning change, the real-world testing and evidence gathered will be the ultimate arbiter of truth. Adding cluster keys has cost, time, and performance trade-offs to consider too, which I discuss later within this chapter.

A table with a clustering key defined is considered to be clustered.

Snowflake also states that clustering keys are not intended for all tables due to the costs of initially clustering the data and maintaining the clustering.

Clustering may be considered to improve query response times; note that additional maintenance costs will be incurred, but clustering is not valid in all circumstances. Your mileage will vary.

As discussed within the previous chapter, micro-partitions are the storage containers managed by the Snowflake virtual private client (VPC) and hosted within your CSP environment. You know micro-partitions are immutable and new micro-partitions are written when data changes. This chapter focuses on cluster keys declared for tables, dynamic tables, and materialized views.

You can find more information at `https://docs.snowflake.com/en/user-guide/tables-clustering-keys`.

Facts Relating to Cluster Keys

I will now state some facts relating to cluster keys; this list is not exhaustive but sets out core principles relating to cluster keys.

- Cluster keys are not micro-partitions; they exist to co-locate data within micro-partitions.

- A cluster key is not the same as a primary key.

- A table can have only one cluster key.

- Alternate cluster keys are implemented by materialized views (MVs).

- Dynamic tables (DTs) *cannot* have a cluster key.

- A cluster key should not exceed five or six composite attributes; there is, preferably, a maximum of four. Adding more than three to four columns tends to increase costs more than benefits.

- Cluster key attributes must be ordered from:

  - Lowest cardinality first

  - Highest cardinality last

- Cluster key attributes must match query predicates.

- Cluster keys are used for micro-partition pruning.

  - The leading attribute with the lowest cardinality prunes the highest number of micro-partitions.

  - Attributes with higher cardinality prune fewer micro-partitions.

You can find more information on cluster keys at `https://docs.snowflake.com/en/user-guide/tables-clustering-keys`.

Cluster Keys and Unique Indexes

Having stated some facts relating to cluster keys, let's now consider how cluster keys differ from primary keys and unique indexes. There is a common misconception that a cluster key is the same as, or analogous to, a primary key or composite unique index, but they are not similar. Here I explain the differences.

In an online transaction processing (OLTP) database, you expect transactions to be stored in tables with varying frequency. You have no control over the velocity, volume, and variety of transactions, just that your database is expected to efficiently process and store data for later retrieval. OLTPs are optimized for storage where inbound data is typically stored as it arrives, resulting in data scattered across the physical storage device where the target object is defined. In most cases, the physical storage location of data loaded into a table does not matter. Indexes are declared and maintained as separate objects holding either pointers to the target records or actual record values. OLTP databases use query predicates to match index entries and then efficiently identify target record locations for retrieval and onward processing. This process is called *index lookup*, and you may declare many indexes to match your data access paths as defined in the query predicates.

In contrast, Snowflake as a data warehouse (DW) processes data very differently. Snowflake is more closely aligned with online analytical processing (OLAP), consuming less volatile data sets in bulk. In other words, OLTP systems often have an OLAP or DW capability added on as the product matures, whereas Snowflake relies upon existing, curated datasets and is built from the ground up as an analytics platform.

You know Snowflake implements immutable storage in the form of micro-partitions; I discussed this approach in Chapter 4. You also know from Chapter 1 that Snowflake stores metadata for each micro-partition, which includes the maximum and minimum value ranges for each attribute within each micro-partition. This information is always up-to-date; as each micro-partition is written, the corresponding metadata is updated and stored.

A cluster key defines the physical order in which data is written into each micro-partition. Because the metadata is always up-to-date, the Snowflake optimizer always knows the exact micro-partitions to access *where the query predicates match*.

The distinction between cluster keys and indexes is important.

In OLTP systems you will see these issues:

- Where the query predicates do not match an index, poor performance can result.

- Index maintenance occurs for every Data Manipulation Language (DML) operation and can be expensive.

- Every index consumes storage, sometimes more than their source table.

- Indexes may contain pointers to data but may also contain data to satisfy query result sets without accessing their underlying objects.

- Constraints and referential integrity must be honored where declared.

- Performance tuning OLTP systems is time-consuming and expensive.

- Sometimes, OLTP optimizers "plan flip," causing performance issues. A plan flip is where the data access path has been tuned but some underlying condition causes the optimizer to choose an alternative execution path.

In Snowflake, you will see these issues:

- An object can have only a single cluster key.

- Where the query predicates do not match the cluster key, poor performance can result.

- You may use MVs with alternate cluster keys, requiring additional storage, micro-partition maintenance, and associated serverless compute cost.

- DTs in common with MVs require additional storage, micro-partition maintenance and associated serverless compute cost.

- Writing immutable micro-partitions is expensive; high DML workloads may cause blocking and processing failures.

- Cluster keys store data sorted in attribute-declared order, leading all other unsorted data within micro-partitions.

- Constraints and referential integrity are not honored where declared but still perform a valuable service by informing a third-party tooling relationship discovery capability.

While Snowflake allows the declaration of primary, unique, nonunique indexes and referential integrity, these are not enforced but exist to do the following:

- Support data discovery by connecting third-party tooling and end users

- Eliminate redundant joins; I discuss this later within this chapter

Figure 5-6 illustrates the difference between traditional OLTP indexing on the left and Snowflake cluster key explicit data ordering within micro-partitions.

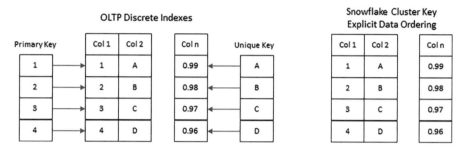

Figure 5-6. *OLTP index and Snowflake cluster key comparison*

You should not expect OLTP workloads to provide comparable performance when ported "like for like" to Snowflake.

You must tune your code for Snowflake and not make assumptions.

Snowflake uses cluster keys for efficient micro-partition pruning; you must be mindful of how the Snowflake optimizer functions to maximize performance. I discuss warehouse sizing and the query performance impact in a later chapter.

Logical Structure and Physical Storage

The logical structure of a cluster key represents the data set layout returned to the user interface as a consequence of issuing a query. The physical storage represents the data set as physically stored on disk. To illustrate the difference, let's investigate the region_baseline data, deliberately chosen as there is a single micro-partition and very small data set.

```
SELECT r_name,
       r_regionkey,
       r_comment
FROM   region_baseline;
```

You expect an unordered dataset to be returned; however, you will see an ordered list and suggest the data at the point of source table creation was explicitly ordered. Ordered inserts are discussed later within this chapter.

- Physical Storage on Disk shows the probable contiguous serial (block device) nature of information stored within a micro-partition.

- Ordered Record Storage shows each physical record with the end of record interpreted as a carriage return/line feed (CRLF).

- Logical Query Result Set shows the record set returned from the query according to attribute order selected.

Figure 5-7 illustrates the three representations of the selected data set. You may also assume each attribute has a terminator value appended to mark the end of the data.

Physical Storage on Disk

0	AFRICA	lar deposits…	1	AMERICA	hs use ironic…	2	ASIA	ges. thinly…	3	EUROPE	ly final courts…	4	MIDDLE EAST	uickly special…

Ordered Record Storage on Disk

0	AFRICA	lar deposits…
1	AMERICA	hs use ironic…
2	ASIA	ges. thinly…
3	EUROPE	ly final courts…
4	MIDDLE EAST	uickly special…

Logical Query Result Set

R_NAME	R_REGIONKEY	R_COMMENT
AFRICA	0	lar deposits. b
AMERICA	1	hs use ironic, ‹
ASIA	2	ges. thinly eve
EUROPE	3	ly final courts
MIDDLE EAST	4	uickly special

Figure 5-7. Storage and display formats

Cluster Key Management

In this section, I will discuss how to manage and maintain the table content specifically with regard to cluster key management. I will now discuss operational considerations for cluster key management:

- Cluster keys should be defined only for high-frequency queries with matching query predicates.

- Cluster key maintenance may not happen concurrently with DML operation completion; maintenance requires a finite time to complete.

- Frequent DML operations may result in costly reclustering operations and, in worst-case scenarios, constant micro-partition "churn."

- Clustering is most cost effective for low-volume DML and high-volume query operations.

- Reclustering invalidates cached results.

The evidence points to tuning your design by considering the nature of DML operations and the likely impact upon micro-partition and cluster key maintenance.

I will now explain how to manage cluster keys.

Investigating Unclustered Tables

At this point you may be asking yourself what happens when data is loaded into a table without a defined cluster key. When you created your baseline tables, you copied the source TPC tables and produced baseline tables without an explicitly defined cluster key. This is an exercise you will now repeat to investigate the natural cluster key for a new table.

Unfortunately, you are not able to view natural cluster keys directly. That is, there is no command available to extract cluster key information from an unclustered table. To prove this assertion, create an unclustered table called `lineitem_unclustered`.

```
USE WAREHOUSE IDENTIFIER ( $tpc_warehouse_xl );


CREATE OR REPLACE  TABLE lineitem_unclustered
AS
SELECT l_orderkey,    l_partkey,      l_suppkey, l_linenumber,
       l_quantity,    l_extendedprice, l_discount, l_tax,
       l_returnflag,  l_linestatus,    l_shipdate, l_commitdate,
       l_receiptdate, l_shipinstruct,  l_shipmode, l_comment
FROM   snowflake_sample_data.tpch_sf1000.lineitem;
```

Revert to an X-Small warehouse:

```
USE WAREHOUSE IDENTIFIER ( $tpc_warehouse_xs );
```

Prove your table is unclustered. You should see this error message: "000005 (XX000): Invalid clustering keys or table LINEITEM_UNCLUSTERED is not clustered."

```
SELECT parse_json(system$clustering_information ( 'lineitem_
unclustered' ));
```

You'll now see how default clustering is applied. From prior experience, you know that an unclustered table requires an attribute to be passed, but this has unforeseen consequences.

For any table, `system$clustering_information` returns clustering information for the input attributes regardless of the declared cluster key.

Understanding how `system$clustering_information` works is important.

To illustrate your assertion, consider these two checks against your unclustered table `lineitem_unclustered`:

```
SELECT parse_json(system$clustering_information ( 'lineitem_unclustered',
'(l_shipdate)' ));
```

```
SELECT parse_json(system$clustering_information ( 'lineitem_unclustered',
'(l_comment)'  ));
```

In Figure 5-8, you can see two distinctly different responses from the previous commands.

```
{
  "average_depth": 2282.4575,
  "average_overlaps": 3753.3499,
  "cluster_by_keys": "LINEAR(l_shipdate)",
  "clustering_errors": [],
  "partition_depth_histogram": {
    "00000": 0,
    "00001": 70,
    "00002": 0,
    "00003": 0,
    "00004": 0,
    ...
    "00128": 4,
    "00256": 14,
    "00512": 57,
    "01024": 579,
    "02048": 1361,
    "04096": 7312
  },
  "total_constant_partition_count": 80,
  "total_partition_count": 9398
}
```

```
{
  "average_depth": 9393,
  "average_overlaps": 9392,
  "cluster_by_keys": "LINEAR(l_comment)",
  "clustering_errors": [],
  "notes": "Clustering key columns contain high cardinality key
  "partition_depth_histogram": {
    "00000": 0,
    "00001": 0,
    ...
    "00014": 0,
    "00015": 0,
    "00016": 0,
    "16384": 9393
  },
  "total_constant_partition_count": 0,
  "total_partition_count": 9393
```

Figure 5-8. *Unclustered table information*

Note the total_partition_count value differs due to the compression algorithm being applied to two different leading cluster key attributes; this is the expected behavior. Later in this chapter you will use system$clustering_information to predict clustering where your predicted value matches your real-world value.

To close out your investigation and check your findings, let's check the clustering depth using the same attributes.

```
SELECT parse_json(system$clustering_depth ( 'lineitem_unclustered',
'(l_shipdate)' ));
```

```
SELECT parse_json(system$clustering_depth ( 'lineitem_unclustered',
'(l_comment)' ));
```

You should see two different values, 2293.3921 and 9393, confirming your assertion is correct.

Use system$clustering_information before and after making changes to the cluster keys as part of your performance tuning process.

I have proven it is not possible to infer how natural cluster keys are defined by using `system$clustering_information` to probe an unclustered table. The only option remaining is to issue queries against an unclustered table with predicates designed to invoke pruning, an exercise I leave for your further investigation.

Default Clustering on Data Load

A cluster key affects the attributes and attribute data content storage order within each micro-partition.

Regardless of whether a cluster key is explicitly declared, all micro-partitions have a cluster key. The default, nonexplicitly declared cluster key is derived from the data set attribute insert order. This is called the *natural cluster key*.

There are several ways to influence the natural cluster key definition.

- Explicitly order the attributes within the source file before load

- From an existing table use `INSERT` with `ORDER BY` to explicitly order the attributes

- Create a view over the existing table with an embedded `ORDER BY`

I do not make any recommendation for ordering data as part of the `INSERT` operation. Simply stating this approach may have merit particularly for tables with very low DML activity. Your approach will be determined by the data refresh frequency and any subsequent DML operations.

The `COPY` command does not allow attribute ordering. If explicit attribute ordering is required, use a staging table to pre-load and then an ordered `INSERT` or `VIEW`.

For many use cases, the natural cluster key is sufficiently performant and delivers adequate pruning to not require micro-partition maintenance. Of course, the initial attribute insert order is significant in deriving the natural cluster key, and DML activity may lead to later data skewing. For relatively static datasets that have been explicitly ordered on insert and assuming the data access paths are also well known, you can expect ordered loads to provide adequate performance.

When manually declaring a cluster key, you must be mindful of factors that influence your decisions.

Attribute Cardinality

Micro-partition pruning is dependent upon the cluster key attribute ordering. When considering cluster key candidates, you must investigate the cardinality of each attribute.

Tables typically contain these types of data:

- Primary and foreign keys derived from sequences

- Natural keys and composite unique keys that may repeat across records

- Textual descriptive attributes including flags, dates, and timestamps

- Volatile attributes whose values are expected to differ for each row

- Semi-structured data such as JSON, AVRO, Parquet, and ORC

As a rule of thumb, focus on the static elements of your target table as the volatile attributes will certainly have high cardinality.

Numeric, partial strings, and partial timestamps represent the best candidates for deriving low-cardinality attributes suitable for cluster keys.

The following list represents the order of preference when determining cluster key attributes:

- Numeric data types are preferred over all other data types.

- Dates/timestamps are represented internally as numeric data types.

- Dates are preferred over timestamps; better still, use YYYYMM.

- The first five to six characters of a string are good candidates too.

These data types are not good candidates for cluster key attributes:

- Geography, variant, and array are not supported in general.

- Some JSON elements are supported.

Cluster Key Lifecycle

Cluster keys can be applied to tables, dynamic tables, and materialized views. Cluster keys relate to the columnar ordering of data within micro-partitions.

Cluster keys are implemented by the following:

- Implicit definition at data load time

- Explicit declaration at object creation time

- Later application by altering an existing object

- Removal by altering an existing object

As identified earlier within this chapter, you must judiciously use cluster keys; there are no silver bullets.

Investigating a Cluster Key

In addition to their use in identifying micro-partition counts for objects, you can use the Snowflake-supplied functions system$clustering_depth and system$clustering_information to investigate cluster keys.

I assume your execution context has been set up, after which you can check clustering information for an unclustered table:

```
SELECT parse_json(system$clustering_information ( 'lineitem_baseline',
'(l_shipdate)' ));
```

Figure 5-9 illustrates partial sample JSON output.

PARSE_JSON(SYSTEM$CLUSTERING_INFORMATION ('LINEITEM_BASELINE', '(L_SHIPDATE)'))

{ "average_depth": 1410.8451,   "average_overlaps": 2197.0682,   "cluster_by_keys": "LINEAR(l_shipdate)",

Figure 5-9. *Example cluster key information*

By using https://jsonformatter.org/ to reformat the JSON results and removing repeating content, Figure 5-10 shows clustering information for lineitem_baseline.l_shipdate using the same query shown in Figure 5-9.

```
{
  "average_depth": 1410.8451,
  "average_overlaps": 2197.0682,
  "cluster_by_keys": "LINEAR(l_shipdate)",
  "clustering_errors": [],
  "partition_depth_histogram": {
    "00000": 0,
    "00001": 147,
    "00002": 0,
    ......
    "00512": 274,
    "01024": 978,
    "02048": 7966
  },
  "total_constant_partition_count": 158,
  "total_partition_count": 9410
}
```

Figure 5-10. *Example cluster key expanded information*

The following are the important attribute values to note.

average_depth

This is the average cluster depth defined. The `average_depth` calculation is derived from the depths recorded for micro-partition values at each level and therefore is a relative value for comparison purposes. You cannot derive absolute information using averages as you do not know the algorithm used to derive the resultant number. You therefore use `average_depth` as an indicator of "good" or "bad" where 1 is your target value.

average_overlaps

This is the average cluster width defined earlier. The `average_overlaps` calculation is derived from the number of micro-partitions where content overlaps with others and is a relative value for comparison purposes. You cannot derive absolute information using averages as you do not know the algorithm used to derive the resultant number. You therefore use `average_overlaps` as an indicator of "good" or "bad" where 1 is your target value.

total_partition_count

This is the object micro-partition count, which may vary according to the compression of cluster key attributes. You should expect minor variations in micro-partition counts when clustering tables using different attributes and ordering due to the compression algorithm.

total_constant_partition_count

This value shows the number of micro-partitions for which the attribute l_shipdate values have reached an optimal clustering state, indicating the micro-partitions will not benefit significantly from reclustering. In this example, total_constant_partition_count is 158, a relatively low value when compared to the total_partition_count value of 9410. You should aim for total_constant_partition_count to be as close to total_partition_count as possible.

As your table is unclustered, i.e., no cluster key has been explicitly declared, a low value for total_constant_partition_count provides some insight into the default clustering key defined upon initial data load.

Within partition_depth_histogram, you see two value ranges. On the left you have the cluster depth, and on the right, you have the number of micro-partitions at the specified cluster depth. Figure 5-11 shows the partial partition depth histogram.

```
"partition_depth_histogram": {
    "00000": 0,
    "00001": 147,
    "00002": 0,
    ......
    "00512": 274,
    "01024": 978,
    "02048": 7966
},
```

Cluster Depth

Micro-Partition Count

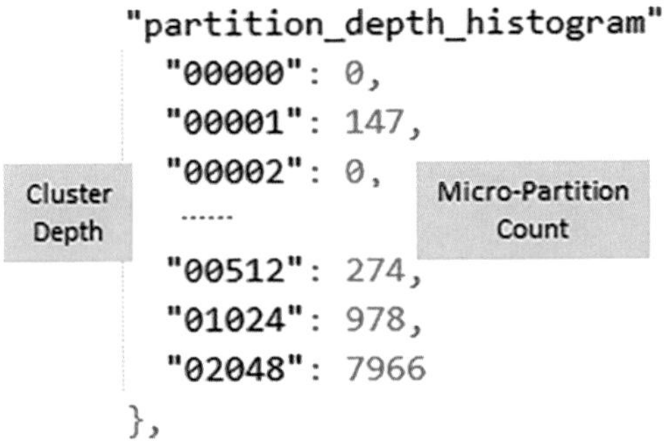

Figure 5-11. *Partition depth histogram*

What does Figure 5-11 tell you?

Using dummy values, Figure 5-12 illustrates how you might imagine micro-partitions are distributed at each depth.

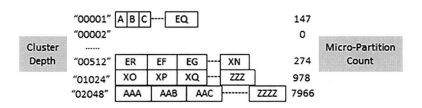

Figure 5-12. *Partition depth micro-partition distribution*

With these metrics you are able to investigate the appropriate cluster key attributes subject to the matching query predicates.

You know that the `lineitem_baseline.l_shipdate` data type is declared as a `DATE` and therefore has no time component. In a real-world scenario you would expect new records to be appended on a regular basis; therefore, your table will grow, and cardinality will increase slightly each business day. If you are to consider `l_shipdate` as a viable candidate for inclusion within a cluster key, you must identify the existing data cardinality.

```
SELECT COUNT ( DISTINCT l_shipdate )
FROM   lineitem_baseline;
```

You expect 2,426 distinct values to be returned.

As previously discussed, low cardinality is preferred for cluster key leading attributes, and 2,426 is not sufficiently low enough for consideration, particularly as the cardinality will increase each business day. But there is a way to reduce the cardinality of `l_shipdate` by using year and month only.

```
SELECT COUNT ( DISTINCT DATE_PART ( YEAR,  l_shipdate )||
                        DATE_PART ( MONTH, l_shipdate ))
FROM   lineitem_baseline;
```

You expect 84 distinct values to be returned with the abbreviated format of "YYYYMM," which is sufficiently low cardinality to be considered a constituent part of a cluster key. Note that query predicates will need to match the abbreviated year/month format. Snowflake recommends abbreviating dates to "YYYYMM" format to reduce cardinality for cluster key attributes.

Using these tools, you are able to investigate candidate attributes for cluster keys, an exercise I will leave to you to further investigate.

Good and Bad Partition Depth Histograms

With an understanding of cluster depth, you can understand what constitutes "good" and "bad." In other words, if you understand what "good" looks like, then as you iterate through progressive improvements with your clustering strategy, you have a goal to shoot for. In programming we often look for patterns; if code is well-formatted, it is often an indicator of well-written code. Likewise, with partition depth histograms, if you know what "good" looks like, then "bad" or "not so good" will be self-evident.

Performance tuning often reaches a point of diminishing returns where further investment in time and effort yields little further benefit. Remembering that cluster keys are but one technique in your toolkit of available tuning tools, knowing when to stop tuning is a valuable skill to develop. With this in mind, Figure 5-13 shows a "good" or "half-decent" partition depth histogram.

```
"partition_depth_histogram": {
    "00000": 0,
    "00001": 8310,
    "00002": 599,
    "00003": 844,
    "00004": 417,
    "00005": 149,
    "00006": 17
}
```

Cluster Depth

Micro-Partition Count

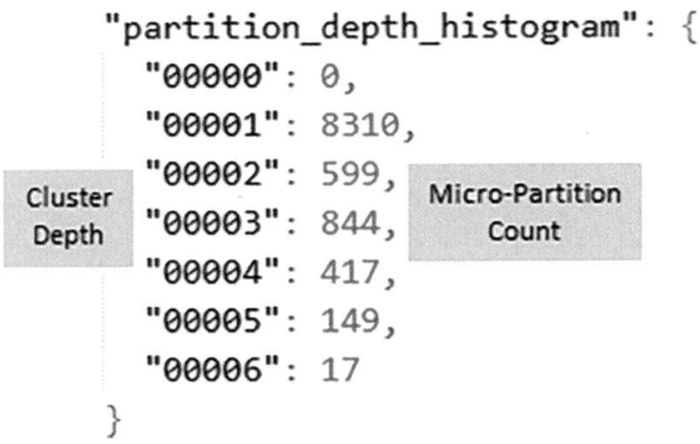

Figure 5-13. *"Good" or "half-decent" partition depth histogram*

Why does the partition depth histogram in Figure 5-13 represent a "good" or "half-decent" example?

To answer the question, let's look at the cluster depth on the left and micro-partition count on the right. You are looking for the following:

- A low incremental count for cluster depth, in this example "00001" to "00006" indicating overall low cluster depth.

- Decreasing values for micro-partition count, in this example 8310 to 17 indicating that approximately 80 percent (8310) of micro-partitions do not overlap.

Figure 5-14 shows a partition depth histogram where every micro-partition except for those four at cluster depth "00001" overlaps with at least one other.

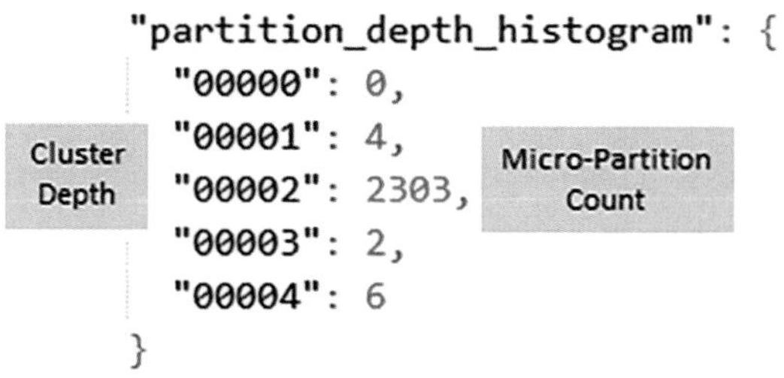

```
"partition_depth_histogram": {
    "00000": 0,
    "00001": 4,
    "00002": 2303,
    "00003": 2,
    "00004": 6
}
```

Cluster Depth

Micro-Partition Count

Figure 5-14. *"Poor" partition depth histogram*

A poor clustering key will always prompt a "notes" attribute within the JSON message, as shown in Figure 5-15.

```
"notes" : "Clustering key columns contain high cardinality key PS_SUPPKEY which might result in
   expensive re-clustering. Consider reducing the cardinality of clustering keys. Please refer to https
   ://docs.snowflake.net/manuals/user-guide/tables-clustering-keys.html for more information.",
```

Figure 5-15. *"Poor" cluster key warning message*

Let's define a cluster key for real!

Defining a Cluster Key

The primary purpose of a cluster key is to enable efficient micro-partition pruning. You must ensure your leading cluster key attribute has the lowest cardinality and matches your most commonly used query predicates. In the previous section, you determined how to abbreviate a date attribute to reduce cardinality and, when included within a cluster key, improve pruning.

Using information derived from the previous section, you can state some objectives:

- average_depth: To be as close to 1 as possible

- average_overlaps: To be as close to 1 as possible

- total_constant_partition_count: To be as close to total_partition_count as possible

I now illustrate how a table with an abbreviated cluster key is created at object creation time; note the use of an X-Large warehouse.

```
USE WAREHOUSE IDENTIFIER ( $tpc_warehouse_xl );
```

Create a view with an additional abbreviated attribute l_shipdate_yyyy_mm for later use when defining the clustered table.

```
CREATE OR REPLACE VIEW v_lineitem_clustered
AS
SELECT l_orderkey,   l_partkey,       l_suppkey,  l_linenumber,
       l_quantity,   l_extendedprice, l_discount, l_tax,
       l_returnflag, l_linestatus,    l_shipdate, l_commitdate,
       l_receiptdate, l_shipinstruct,  l_shipmode, l_comment,
       DATE_PART ( YEAR,  l_shipdate )||
           DATE_PART ( MONTH, l_shipdate ) AS l_shipdate_yyyymm
FROM   lineitem_baseline;
```

Create a clustered table from the view v_lineitem_clustered with an ordered insert. If you had declared your source view v_lineitem_clustered with an ORDER BY, then every invocation would force a sort, and in real-world use, you probably would not want to implicitly order every query. You can add an ORDER BY clause to your Create Table AS (CTAS) statement to force ordering at the point of data load.

Note that the creation of your clustered table with ordered data takes about 8½ minutes.

```
CREATE OR REPLACE TABLE lineitem_clustered
CLUSTER BY ( l_shipdate_yyyymm )
AS
SELECT l_orderkey,   l_partkey,       l_suppkey,  l_linenumber,
       l_quantity,   l_extendedprice, l_discount, l_tax,
       l_returnflag, l_linestatus,    l_shipdate, l_commitdate,
       l_receiptdate, l_shipinstruct,  l_shipmode, l_comment,
       l_shipdate_yyyymm
FROM   v_lineitem_clustered
ORDER BY l_shipdate_yyyymm;
```

Let's now define a cluster key for lineitem_baseline using your abbreviated date for l_shipdate; then examine the clustering effect.

```
USE WAREHOUSE IDENTIFIER ( $tpc_warehouse_xs );

ALTER TABLE lineitem_baseline
```

```
CLUSTER BY ( DATE_PART ( YEAR,  l_shipdate )||
             DATE_PART ( MONTH, l_shipdate ));
```

Note the `ALTER TABLE` command completes almost instantaneously, but the clustering activity occurs asynchronously "behind the scenes."

Automatic clustering is enabled by default when a table is reclustered and uses serverless compute; I discuss automatic clustering later in the chapter.

Creating a cluster key means you no longer need to specify an attribute when selecting clustering information.

```
SELECT parse_json(system$clustering_information ( 'lineitem_baseline' ));
```

Figure 5-16 shows the resultant JSON.

```
{
    "average_depth": 6211.5008,
    "average_overlaps": 6658.2023,
    "cluster_by_keys": "LINEAR( DATE_PART ( YEAR,l_shipdate )||DATE_PART ( MONTH,l_shipdate ))",
    "clustering_errors": [],
    "partition_depth_histogram": {
      "00000": 0,
      "00001": 1477,           Micro-Partition
      ......                      Count
      "08192": 7943
    },
    "total_constant_partition_count": 1491,
    "total_partition_count": 9420
}
```

Cluster Depth

Figure 5-16. *Partial date cluster key defined*

Figure 5-17 compares "before" and "after." On the left it shows the original unclustered micro-partition representation; on the right it shows your newly clustered representation.

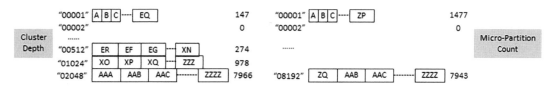

Figure 5-17. *Unclustered and clustered table comparison*

From a clustering perspective only, you might consider the addition of your single attribute cluster key is a decent first step toward delivering high selectivity. Figure 5-17 shows an increase in cluster depth micro-partition count from 147 to 1477 indicating more micro-partitions would be pruned when the query predicates match your cluster key, and all other micro-partitions have been relegated to a much deeper depth: 2048 to 8196.

You can test the effectiveness of your cluster key by performing before and after queries.

```
USE WAREHOUSE IDENTIFIER ( $tpc_warehouse_m );
```

Disable the local cache.

```
ALTER SESSION SET use_cached_result = FALSE;
```

You can check the warehouse status via the output of the following:

```
SHOW WAREHOUSES LIKE '<warehouse_name>';
```

Suspending and resuming warehouses clears the local cache; however, sometimes the SUSPEND command fails as the warehouse is not running. Simply RESUME and then SUSPEND.

```
ALTER WAREHOUSE IDENTIFIER ( $tpc_warehouse_m ) SUSPEND;
ALTER WAREHOUSE IDENTIFIER ( $tpc_warehouse_m ) RESUME;
```

You do not expect the metadata to contain metrics for partial attribute selection.

```
SELECT COUNT(1)
FROM   lineitem_baseline
WHERE  DATE_PART ( YEAR,  l_shipdate )||
       DATE_PART ( MONTH, l_shipdate ) = '199610';
```

The query should return 77,301,405 rows.

Click the Query ID to examine partition pruning; you should see "Partitions scanned" = 2631.

Instead of removing the cluster key, you re-create your lineitem_baseline table; you do this to remove the possibility of the following:

- Reclustering service not activating as the table is already clustered and not fragmented

- Reclustering service activates and determines a recluster is not needed

- Asynchronous reclustering not activating and recluster completing within your testing timeframe

You simply cannot be sure of how automatic reclustering is triggered or when the reclustering service runs.

```
USE WAREHOUSE IDENTIFIER ( $tpc_warehouse_xl );

CREATE OR REPLACE TABLE lineitem_baseline
AS
SELECT l_orderkey,    l_partkey,      l_suppkey,  l_linenumber,
       l_quantity,    l_extendedprice, l_discount, l_tax,
       l_returnflag,  l_linestatus,   l_shipdate, l_commitdate,
       l_receiptdate, l_shipinstruct,  l_shipmode, l_comment
FROM snowflake_sample_data.tpch_sf1000.lineitem;

USE WAREHOUSE IDENTIFIER ( $tpc_warehouse_m );
```

Disable the local cache.

```
ALTER SESSION SET use_cached_result = FALSE;
```

One can check the warehouse status via output of the following:

```
SHOW WAREHOUSES LIKE '<warehouse_name>';
```

Suspending and resuming warehouses clears the local cache; however, sometimes the SUSPEND command fails as the warehouse is not running. Simply RESUME and then SUSPEND.

```
ALTER WAREHOUSE IDENTIFIER ( $tpc_warehouse_m ) SUSPEND;
ALTER WAREHOUSE IDENTIFIER ( $tpc_warehouse_m ) RESUME;

SELECT COUNT(1)
FROM   lineitem_baseline
WHERE  DATE_PART ( YEAR,  l_shipdate )||
       DATE_PART ( MONTH, l_shipdate ) = '199610';
```

Figure 5-18 shows the difference between a clustered and unclustered table where efficient micro-partitioning is evidenced by the number of "Partitions scanned."

Clustered		Unclustered	
Partitions scanned	2631	Partitions scanned	8483
Partitions total	9625	Partitions total	9406

Figure 5-18. *Partition scan comparison*

To complete your cluster key investigation, you can refer to Figure 5-16, repeated here as Figure 5-19 for reference.

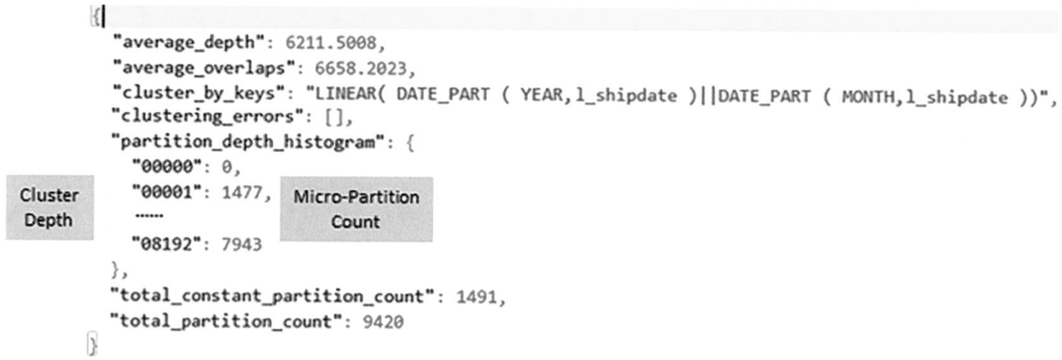

Figure 5-19. *Partial date cluster key defined (same as Figure 5-16)*

In Figure 5-19 you can observe that both `average_depth` and `average_overlaps` have increased. From the evidence available, you can assume the answer relates to these metrics being averages and all micro-partitions are at either cluster depth 00001 or 08192. The observed values for both `average_depth` and `average_overlaps` may not indicate poor micro-partition clustering as the cluster depth has increased.

You also observe metric `total_constant_partition_count` has increased to 1491, a much-improved value when compared to both the unclustered value of 158 and the `total_partition_count` value of 9410.

Lastly, note the small difference in the `total_partition_count` value; this may be due to a more efficient compression of cluster key attributes.

Determining an effective cluster key for a known set of query predicates is relatively simple but requires time and effort to achieve an optimal outcome. Cluster key creation is not a one-step activity and must be considered in the wider context of being a single performance optimization among several available. Alternatively, adding cluster keys may not offer any performance improvement at all; test using "real-world" scenarios before submitting code for release.

Cluster keys alone are not a silver bullet. They are part of an overall performance tuning strategy.

Within this section I have outlined the steps used to determine whether a single abbreviated attribute is suitable for inclusion within a cluster key.

Alternative Cluster Keys

As a general principle, you should take every opportunity to enrich your data by adding attributes to facilitate queries by expected query predicates.

It is not usual to add attributes at the point of ingestion into staging (or raw) tables as you want to preserve parity with the originating data source. As part of your data curation process, where you often SELECT data to create your data products, you may also enrich data by adding attributes to facilitate subsequent consumption.

In this "left-to-right ingest, curate, consume" pattern, you will often create tables, secure views, secure functions, materialized views, and dynamic tables. You know a table can have only a single cluster key, which should be focused on the primary data access path query predicates. If you have multiple data access paths, you face a dilemma: how do you deliver optimal performance when only a single cluster key exists and legacy indexing options are not supported?

You have options: both materialized view and dynamic tables implement their own cluster keys, which are dependent upon a parent table content but not tied to the parent table cluster key.

In this section, I discuss both materialized view and dynamic tables from two perspectives.

- Providing an alternate cluster key

- Enriching your data by adding attributes to improve cluster key selectivity

When the underlying table contents change, so do dependent materialized views and dynamic table contents. Micro-partition reclustering or consolidation causes cached result sets to be invalidated, preventing reuse.

Cached result set validity is subject to dependent micro-partitions remaining untouched.

I will discuss materialized view and dynamic tables next where you retain the same attributes and structure to facilitate cluster key comparison only. Your real-world experience will differ when using both materialized views and dynamic tables.

Materialized Views

In this section, you will create an unclustered materialized view called mv_lineitem_ baseline from a sample TPC table and derive a new partial date attribute called l_ shipdate_yyyymm; note that this step will take a few minutes to complete.

In this example, I have selected all the available attributes for consistency throughout this chapter. In real-world usage, you would expect materialized views to target specific workloads and contain an appropriate subset of attributes along with filters, summaries, and aggregations.

You must remain mindful of the storage implications and reclustering costs when using materialized views.

```
USE WAREHOUSE IDENTIFIER ( $tpc_warehouse_xl );

CREATE OR REPLACE MATERIALIZED VIEW mv_lineitem_baseline
COPY GRANTS
AS
SELECT l_orderkey,    l_partkey,      l_suppkey,  l_linenumber,
       l_quantity,    l_extendedprice, l_discount, l_tax,
       l_returnflag,  l_linestatus,    l_shipdate, l_commitdate,
       l_receiptdate, l_shipinstruct,  l_shipmode, l_comment,
       DATE_PART ( YEAR,  l_shipdate )||
          DATE_PART ( MONTH, l_shipdate ) AS l_shipdate_yyyymm
FROM   snowflake_sample_data.tpch_sf1000.lineitem;

USE WAREHOUSE IDENTIFIER ( $tpc_warehouse_xs );
```

Then check whether the materialized view is clustered.

```
SELECT parse_json(system$clustering_information ( 'mv_lineitem_
baseline' ));
```

You should see this error message: *"000005 (XX000): Invalid clustering keys or table MV_LINEITEM_BASELINE is not clustered."*

Let's check expected clustering using the new attribute l_shipdate_yyyymm.

```
SELECT parse_json(system$clustering_information ( 'mv_lineitem_baseline',
'(l_shipdate_yyyymm)' ));
```

Figure 5-20 shows the expected cluster map using the l_shipdate_yyyymm attribute.

```
{
  "average_depth": 1889.6567,
  "average_overlaps": 2944.7983,
  "cluster_by_keys": "LINEAR(l_shipdate_yyyymm)",
  "clustering_errors": [],
  "partition_depth_histogram": {
    "00000": 0,
    "00001": 1580.
    ......
    "00512": 64,
    "01024": 333,
    "02048": 1719,
    "04096": 5641
  },
  "total_constant_partition_count": 1601,
  "total_partition_count": 9337
}
```

Figure 5-20. *Materialized view l_shipdate_yyyymm cluster key defined*

Now cluster the materialized view mv_lineitem_baseline.

```
ALTER MATERIALIZED VIEW mv_lineitem_baseline
CLUSTER BY ( l_shipdate_yyyymm );
```

Then check that the actual clustering matches the expected clustering.

```
SELECT parse_json(system$clustering_information ( 'mv_lineitem_
baseline' ));
```

You expect clustering results to match those shown in Figure 5-20, which confirms the earlier analysis of how `system$clustering_information` works.

Materialized views have many capabilities beyond providing an alternate access path via a cluster key. You can find more information at `https://docs.snowflake.com/en/user-guide/views-materialized`.

You can find more information on clustering materialized views at `https://docs.snowflake.com/en/user-guide/views-materialized#label-clustering-base-table-and-materialized-view`.

Dynamic Tables

Dynamic tables (DTs) at the time of writing are in public preview status. You can find more information at `https://docs.snowflake.com/en/user-guide/dynamic-tables-about`.

For those with a legacy RDBMS background, dynamic tables fulfil the same capability as using complex or simple materialized views. Snowflake has differentiated, or confused, developers by renaming a familiar concept *dynamic tables*.

After creating the unclustered materialized view `mv_lineitem_baseline`, you now can create a dynamic table equivalent. You use the same sample TPC table when you declare a materialized view called `mv_lineitem_baseline` and derive the same partial date attribute `l_shipdate_yyyymm`.

As with your MV example, you have selected all available attributes for consistency throughout this chapter. In real-world usage, you would expect dynamic tables to target specific workloads and contain an appropriate subset of attributes along with filters, summaries, and aggregations.

You must remain mindful of the storage implications and reclustering costs when using dynamic tables.

```
CREATE OR REPLACE DYNAMIC TABLE dt_lineitem_baseline COPY GRANTS
TARGET_LAG = '30 MINUTES'
WAREHOUSE  = tpc_wh_xsmall
AS
```

```
SELECT l_orderkey,     l_partkey,       l_suppkey,  l_linenumber,
       l_quantity,     l_extendedprice, l_discount, l_tax,
       l_returnflag,   l_linestatus,    l_shipdate, l_commitdate,
       l_receiptdate,  l_shipinstruct,  l_shipmode, l_comment,
       DATE_PART ( YEAR,  l_shipdate )||
           DATE_PART ( MONTH, l_shipdate ) AS l_shipdate_yyyymm
FROM   snowflake_sample_data.tpch_sf1000.lineitem;
```

Because of referencing an imported shared object, we expect to see this error: "Insufficient privileges to operate on base table to automatically enable CHANGE_TRACKING for dynamic table 'DT_LINEITEM_BASELINE.'"

Here you can see a difference in behavior between materialized views and dynamic tables. You must replace the referenced imported shared view with the local table tpc.tpc_owner.lineitem_baseline.

The initial DT build does not use the DT declared warehouse, which is used only for refreshes.

Regardless of warehouse size, DT creation will take more than four hours; I tested with both X-Small and X-Large warehouses.

```
CREATE OR REPLACE DYNAMIC TABLE dt_lineitem_baseline COPY GRANTS
TARGET_LAG = '30 MINUTES'
WAREHOUSE  = tpc_wh_xsmall
AS
SELECT l_orderkey,     l_partkey,       l_suppkey,  l_linenumber,
       l_quantity,     l_extendedprice, l_discount, l_tax,
       l_returnflag,   l_linestatus,    l_shipdate, l_commitdate,
       l_receiptdate,  l_shipinstruct,  l_shipmode, l_comment,
       DATE_PART ( YEAR,  l_shipdate )||
           DATE_PART ( MONTH, l_shipdate ) AS l_shipdate_yyyymm
FROM   tpc.tpc_owner.lineitem_baseline;
```

As an aside, if you want to change your DT refresh warehouse from X-Small to Small, this command is required:

```
ALTER DYNAMIC TABLE dt_lineitem_baseline
SET WAREHOUSE = tpc_wh_small;
```

Then check whether the dynamic table is clustered:

```
SELECT parse_json(system$clustering_information ( 'dt_lineitem_
baseline' ));
```

You should see this error message: "000005 (XX000): Invalid clustering keys or table DT_LINEITEM_BASELINE is not clustered."

Let's check the expected clustering using your new attribute, l_shipdate_yyyymm.

```
SELECT parse_json(system$clustering_information ( 'dt_lineitem_baseline',
'(l_shipdate_yyyymm)' ));
```

Figure 5-21 shows the abbreviated JSON; note that all the micro-partitions are at a cluster depth of 32768 and total_constant_partition_count = 0, indicating your new attribute l_shipdate_yyyymm is totally ineffective for micro-partition pruning.

```
{
    "average_depth": 18312,
    "average_overlaps": 18311,
    "cluster_by_keys": "LINEAR(l_shipdate_yyyymm)",
    "clustering_errors": [],
    "partition_depth_histogram": {
      "00000": 0,
      "00001": 0,
      ......
      "00016": 0,
      "32768": 18312
    },
    "total_constant_partition_count": 0,
    "total_partition_count": 18312
}
```

Figure 5-21. *Dynamic table l_shipdate_yyyymm no cluster key*

Let's attempt to add a cluster key.

```
ALTER TABLE dt_lineitem_baseline
CLUSTER BY ( l_shipdate_yyyymm );
```

You should receive this error message: "000002 (0A000): Unsupported feature 'Dynamic tables do not support clustering actions.'"

DTs cannot be clustered.

While not essential to your investigation of DTs, here you add RESUME and REFRESH information for completeness.

When the dynamic table is declared, the TARGET_LAG dictates the first runtime; in this example, 30 minutes will elapse before the first refresh.

You must now resume the dynamic table.

```
ALTER DYNAMIC TABLE dt_lineitem_baseline RESUME;
```

You may see this error message: "Dynamic Table 'TPC.TPC_OWNER.DT_LINEITEM_BASELINE' is not initialized. Please run a manual refresh or wait for a scheduled refresh before querying."

Then you must use REFRESH on the dynamic table to force a refresh.

```
ALTER DYNAMIC TABLE dt_lineitem_baseline REFRESH;
```

MV and DT Considerations

Adding materialized views and dynamic tables is not a silver bullet. They are tools within your performance tuning toolkit that may prove beneficial under certain circumstances. These are some factors to consider when determining whether materialized views and dynamic tables are beneficial:

- Consider micro-partition storage when reclustering using MVs and DTs:

  - Filters, aggregates, and summaries reduce data volume.

  - Smaller data volumes may not need a cluster key.

- MVs and DTs should target specific workloads and not be generic in nature.

- MVs can be declared only against a single table, and creation limitations apply: `https://docs.snowflake.com/en/user-guide/views-materialized#limitations-on-creating-materialized-views`.

- MVs do not support window functions and nondeterministic functions, i.e., `current_timestamp()`; a full list of limitations is at `https://docs.snowflake.com/en/user-guide/views-materialized#label-limitations-on-working-with-materialized-views`.

- DT build times can be excessive; use a much larger warehouse to build DTs.

- DTs require change tracking and therefore are not supported for imported shared objects: `https://docs.snowflake.com/en/user-guide/dynamic-tables-tasks-create#dynamic-tables-and-change-tracking`.

- DT replication behavior is not the same between replication groups and failover groups: `https://docs.snowflake.com/en/user-guide/account-replication-considerations#label-replication-and-dynamic-tables`.

- DTs have limitations: `https://docs.snowflake.com/en/user-guide/dynamic-tables-tasks-create#dynamic-table-limitations-and-supported-functions`.

- DML velocity and volume may overwhelm the background clustering process capability.

- Additional processing and storage are required to support each materialized view or dynamic table.

- Query predicates must match cluster key for optimal pruning.

- Adding summary attributes into a cluster key usually improves performance.

- Real-world testing triumphs theory. Test, test, and then test again.

Materialized View Query Rewrite

No discussion on materialized views would be complete without a section on how materialized views can accelerate query performance.

When a query is compiled, query optimization determines whether a materialized view can satisfy part or all of the query at a cheaper cost than the base table. The cheapest execution path is chosen, which may result in the materialized view being chosen instead of the base table.

A much deeper dive into materialized query rewrite is provided by Minzhen Yang at `https://www.linkedin.com/pulse/snowflake-materialized-view-query-auto-rewrite-minzhen-yang/`.

Automatic Clustering

Throughout this chapter I have made reference to automatic clustering, often without explaining the full context. Here I draw together my knowledge into a single section.

The objective for the automatic clustering service is to reduce both cluster depth and cluster width by monitoring the state of clustered tables and reclustering when required. When activated, the automatic clustering service operates behind the scenes, but only for tables where a cluster key has been declared. Manual reclustering has been deprecated for all accounts; therefore, it should not be considered. Reclustering incurs costs, which I discuss later.

Over time your tables may experience fragmentation due to DML activity. Where data has a wide range of low-volume, high-frequency updates, your micro-partitions may be rewritten frequently, a phenomena called *churning*.

Imagine the scenario shown in Figure 5-22 where a subset of micro-partitions is rewritten every 15 minutes.

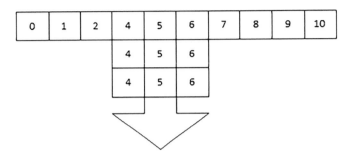

Figure 5-22. *Periodic updates to a subset of micro-partitions*

In this scenario, churning micro-partitions is likely to cause these problems:

- Increased storage costs according to the Time Travel setting, Fail-Safe setting, and cloning activity

- Object locking while the DML completes

- SQL statement queueing until resources become available

- Inability to complete reclustering before the next DML arrives

- Invalidation of cache results: `https://docs.snowflake.com/en/user-guide/querying-persisted-results`

You can find more information on automatic reclustering at `https://docs.snowflake.com/en/user-guide/tables-auto-reclustering`.

Workflow

The automatic clustering workflow is triggered by DML activity making changes to table content, resulting in new micro-partitions.

Let's assume you have a table with clustering enabled. For this table, an `UPDATE` statement changes the content of several micro-partitions resulting in all affected micro-partitions being replaced by new micro-partitions containing new content.

When the transaction completes, the new object content is immediately available. Without investigating the clustering, you cannot tell how well the object is clustered according to its attributes.

At some point, the automatic clustering service initiates, scans the micro-partitions, and determines whether reclustering is required. The automatic clustering service iteratively selects a batch of micro-partitions and reclusters. When complete, the process repeats until it determines further reclustering will not be beneficial.

Reclustering

The automatic clustering service runs asynchronously and uses serverless compute. Credits are consumed by serverless compute, so this process is worth monitoring to ensure excessive resource consumption does not occur. As previously stated, high DML workloads may cause problems where the clustering service cannot complete reclustering optimally before a new workload arrives.

You have also seen how reclustering is not instantaneous; removing an existing cluster key may not trigger a recluster due to the following:

- The reclustering service not activating as the table is already clustered and not fragmented

- The reclustering service activates and determines a recluster is not needed

- The asynchronous reclustering not activating and the recluster completing within your testing timeframe

You simply cannot be sure of how automatic reclustering is triggered or when the reclustering service runs. To determine if automatic clustering is enabled on a table, look for the `automatic_clustering` attribute that is either ON or OFF.

Instead of enabling auto-clustering on all the time and processing delta updates around the clock, some prefer to do this maintenance activity on the weekends so that auto-clustering handles cumulative data changes together.

The compute savings can be significant when suspending reclusters or resuming reclusters before suspending reclusters again. You can find more information at `https://community.snowflake.com/s/article/Periodic-Cleanup-of-Avoidable-Automatic-Clustering-Costs`.

```
SHOW TABLES LIKE 'lineitem_baseline';
```

Automatic clustering is enabled by default when a table is reclustered; you cannot influence automatic clustering behavior except by disabling this:

```
ALTER TABLE lineitem_baseline SUSPEND RECLUSTER;
```

To re-enable automatic clustering, use this:

```
ALTER TABLE lineitem_baseline RESUME RECLUSTER;
```

To determine whether reclustering is complete, check the `average_depth` attribute in the result set for `system$clustering_information`, and when the `average_depth` value remains constant, reclustering is complete.

Cost Monitoring

Monitoring automatic clustering service serverless compute can be achieved in several ways, the first of which is to use the SnowSight console. Navigate to Admin ➤ Usage and then select All Services, Compute, and All Resources, as shown in Figure 5-23.

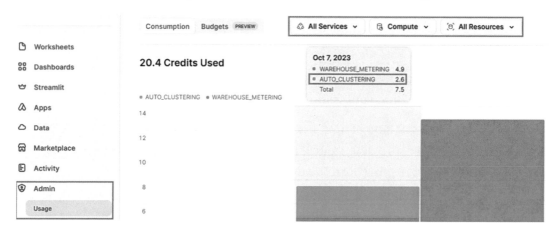

Figure 5-23. *SnowSight consumption*

Consumption information can also be seen by executing this query; note that results are held for only 14 days:

```
SELECT table_name, start_time, end_time, credits_used,
      num_bytes_reclustered,  num_rows_reclustered
FROM TABLE(information_schema.automatic_clustering_history
( date_range_start => DATEADD ( D, -7, current_timestamp )));
```

A corresponding Account Usage Store view is also available. Note that the latency may be up to three hours. You can find more information at `https://docs.snowflake.com/en/sql-reference/functions/automatic_clustering_history`. I will leave this to you to investigate further.

Summary

This chapter began covering the topic of cluster keys by explaining cardinality, cluster width, and cluster depth. I then defined cluster keys, discussed the differences between cluster keys and legacy RDBMS indexes, and then illustrated the physical differences between them.

You discovered that every table has a cluster key whether explicitly declared or implicitly defined by Snowflake at the point of data load. I explained attribute cardinality; from a legacy RDBMS perspective a low-cardinality approach appears counterintuitive. For Snowflake I explained why low-cardinality leading attributes in a cluster key enables higher micro-partition pruning.

Using `system$clustering_information` against an unclustered table revealed how you can use this tool to inform cluster key definition. As you worked through the JSON payload, you discovered average cluster width and average cluster depth are not great values for determining what "good" looks like. Instead, you must look at the ratio of total partitions and constant partitions along with the cluster depth histogram. I also illustrated the value of visual patterns when guiding your investigation. The law of diminishing returns applies to performance tuning for which you can find more information at `https://en.wikipedia.org/wiki/Diminishing_returns`.

Noting a table can have only a single cluster key, you investigated both materialized views and dynamic tables. The investigation revealed that challenges may arise with both materialized views and dynamic tables where high-velocity, low-volume DML transactions occur. You encountered some surprising limitations to dynamic tables too; note that these are in public preview at the time of writing.

Lastly, you investigated automatic clustering to provide both consistent context and usage information. In the next chapter, you will learn about warehouses.

CHAPTER 6

Warehouses

Virtual data warehouses, also called *multicluster compute*, are a first-order named object for a group (or cluster) of servers with multicore/hyperthreading CPU(s), memory, and temporary storage (SSD). Warehouses are massively parallel processing (MPP) engines.

Any running warehouse consumes compute credits; therefore, you should pay careful attention to right-sizing warehouses and their configuration. Not all queries require a running warehouse as some resolve their results directly from metadata or query cache, but for those that do require a running warehouse, if not already done so, the warehouse will instantiate. That is, the cloud services layer will provision the correct number of servers (CPU, memory, and SSD cache) to fulfill the warehouse configuration and begin executing the query.

Warehouses are the single most innovative performance enhancement to data warehousing. Separating compute from storage enables elastic scalability for the most demanding of workloads.

Most tuning effort is directed at resizing warehouses. Many consulting organizations focus on the "low-hanging fruit" of monitoring warehouse consumption and optimizing warehouse size throughout each day's processing, but this approach is flawed for several reasons, which I will discuss in this chapter.

Seemingly a silver bullet for performance tuning workloads, incorrect sizing and use of warehouses leads to both increased cost and poor performance, quite the opposite of our desired outcomes. You must adopt a more informed approach than simply resizing warehouses "on the fly." Warehouse size represents one important dimension for performance tuning. But adopting the blunt instrument of resizing warehouses is not the only consideration.

This chapter has been written from several perspectives so you can:

- Understand how warehouses operate

- Maximize warehouse use

- Reduce warehouse runtime costs

A. Carruthers, *Tuning the Snowflake Data Cloud*, https://doi.org/10.1007/979-8-8688-0379-6_6

- Identify warehouse performance root causes

- Inform application design decisions

- Deliver optimal performance

We cannot achieve all of these objectives without making compromises along the way. My approach is pragmatic and based on real-world experience, including the green-field design, development, and implementation of Snowflake applications; performance tuning of "lift and shift" production applications; replicating data sets globally; and integration with Snowflake Marketplace.

You will often find yourself operating with ever-changing workloads where performance may be fine one day but periodically is unacceptable. Alternatively, often consumers experience inconsistent performance while accessing Snowflake. You must therefore consider performance tuning within the context of the workload executing at the time when performance issues were observed.

Nobody said this was going to be easy. Tracing intermittent issues is problematic. I am reminded of the Observe, Orient, Decide, Act (OODA) Loop coined by Colonel John Boyd of the United States Air Force. Boyd's OODA Loop suggests only 70 to 80 percent of the available information is necessary to make an informed decision.

In fact, waiting for 100 percent of all available information may take too long, so 70 to 80 percent is sufficient for identifying a probable root cause. It's better to attempt an informed fix sooner than later. You can find more information on the OODA loop at `https://en.wikipedia.org/wiki/OODA_loop`.

Identifying Snowflake credit consumption charges for the various services provided is not trivial. You can find a full list of credit consumptions at `https://www.snowflake.com/legal-files/CreditConsumptionTable.pdf`.

I begin this coverage by establishing the foundational information you will need to set the context for your investigation of warehouses before moving on to covering the typical day-to-day workloads performed by your applications.

Foundational Information

In this section I describe the basic concepts relating to warehouses to establish some baseline information that this chapter later relies upon. I do not deep dive into the foundational information but instead provide a brief summary and links to both documentation and other articles for your further investigation. I assume you have some familiarity with the basic principles of memory and compute resources.

I also show how to install and configure tooling in preparation for later use.

Memory and Compute

Traditional computer hardware provisioning, either into our on-prem data center, desktop computer, laptop, or mobile device, is inherently restricted by the number and type of expansion slots available to upgrade additional memory or CPUs. Our data centers and local hardware suffer from further restrictions in the way devices are operated, and interoperability challenges occur across varying operating systems, software versions, and workloads. Ultimately, hardware obsolescence also plays its part where vendors cease support for aging hardware.

The cloud computer physical hardware upgrade path is likewise limited, but with one crucial difference: the way in which each cloud service provider (CSP) provisions service abstracts us from the underlying hardware, spanning multiple devices and removing the limitations of our on-prem data centers.

This is not the full story, but it serves as an abstract explanation sufficient for our purposes. The important point is that CSPs enable elastic provisioning of memory and compute and cater to the failure of both in a seamless manner. Snowflake groups both memory and compute into "warehouses," which are billable units for which further information can be found at `https://docs.snowflake.com/en/user-guide/warehouses-overview`.

Warehouse Types

Snowflake currently supports two types of warehouse.

- Standard warehouses
- Snowpark-optimized warehouses

You are familiar with standard warehouses as these have been the mainstay of Snowflake since platform inception; therefore, I will not describe standard warehouses further.

Snowpark-optimized warehouses are a much more recent addition where additional memory and local cache are provisioned for computation-intensive processes. Snowpark-optimized warehouses default to Medium size and do not support X-Small, Small, X5-Large, and X6-Large sizes. You can find more information on Snowpark-optimized warehouses at `https://www.snowflake.com/blog/snowpark-optimized-warehouses/`.

Workloads should not be mixed across warehouse types; for example, you should use standard warehouses for bulk data ingestion, data curation, and consumption. You should reserve Snowpark-optimized warehouses for use cases where a large memory footprint is required.

For Snowpark-optimized warehouses, limiting concurrency may be preferrable to release resources though fewer concurrent processes will be supported.

To reduce concurrency, this value is applied at the warehouse level and not at an individual cluster within a warehouse. I explain clusters later in this chapter, but for now it is sufficient to know that the following parameter affects all clusters:

```
ALTER WAREHOUSE <warehouse_name> SET MAX_CONCURRENCY_LEVEL = 1;
```

To display the parameter settings for an individual warehouse, use this:

```
SHOW PARAMETERS LIKE 'MAX_CONCURRENCY_LEVEL' IN WAREHOUSE <warehouse_name>;
```

Because of the additional resources provisioned, Snowpark-optimized warehouses' initiation and background maintenance operations may take a little longer than for standard warehouses.

You can find more information at `https://docs.snowflake.com/en/user-guide/warehouses-snowpark-optimized`.

Warehouse Initialization

Snowflake maintains a number of warehouses within each availability zone thereby reducing spin-up time on demand. Through Snowhouse statistical analysis, Snowflake accurately predicts usage, but for new regions where usage patterns are not yet established, predictions are less accurate. Therefore, provisioning delays on the order of 10+ seconds may be experienced. Naturally, as usage patterns evolve, predictions become more accurate.

According to the Snowflake documentation, the initial creation and resumption of a Snowpark-optimized virtual warehouse may take longer than a standard warehouses.

Declaring Warehouses

In this book, you have already declared and used warehouses, which are essential for querying data sets within Snowflake. The following template code creates a standard X-Small warehouse intended for data ingest:

```
USE ROLE sysadmin;

CREATE OR REPLACE WAREHOUSE ingest_xsmall_wh WITH
WAREHOUSE_TYPE      = STANDARD
WAREHOUSE_SIZE      = 'X-SMALL'
AUTO_SUSPEND        = 60
AUTO_RESUME         = TRUE
MIN_CLUSTER_COUNT   = 1
MAX_CLUSTER_COUNT   = 4
SCALING_POLICY      = 'STANDARD'
INITIALLY_SUSPENDED = TRUE;
```

Always set INITIALLY_SUSPENDED = TRUE; otherwise, the warehouse runs for the AUTO_SUSPEND period on declaration.

The following template code creates a Snowpark-optimized Medium warehouse intended for data Snowpark use:

```
CREATE OR REPLACE WAREHOUSE snowpark_opt_medium_wh WITH
WAREHOUSE_TYPE      = 'SNOWPARK-OPTIMIZED'
WAREHOUSE_SIZE      = 'MEDIUM'
AUTO_SUSPEND        = 60
AUTO_RESUME         = TRUE
MIN_CLUSTER_COUNT   = 1
MAX_CLUSTER_COUNT   = 4
SCALING_POLICY      = 'STANDARD'
INITIALLY_SUSPENDED = TRUE;
```

The following shows how to set the warehouse's `auto_suspend` attribute to 60 seconds outside of declaring a new warehouse:

```
SHOW warehouses;
ALTER WAREHOUSE snowpark_opt_medium_wh SET auto_suspend = 60;
```

Every warehouse regardless of size runs for a minimum of 60 seconds, with per-second billing thereafter. You can find more information on warehouses at `https://docs.snowflake.com/en/sql-reference/sql/create-warehouse`.

Using Warehouses

Understanding warehouse consumption is critical to correctly sizing and using warehouses. Once a warehouse has been created, and before you are able to interact with your data, you must declare a warehouse for use, as this example shows:

```
USE WAREHOUSE ingest_xsmall_wh;
```

After declaration, you can run your workload.

Declaring an optimally sized warehouse for the target workload is crucial in reducing costs. But, as you will see, optimal use of warehouses is equally important, and to optimally use warehouses, you must understand workloads.

I discuss concurrency later within this chapter; however, it is important to know that the default maximum concurrency level is eight per cluster. You can reduce this value to increase the available resources to long-running processes, but this may introduce queueing as a consequence.

Warehouse Capacity

Warehouse capacity is expressed in clusters. A single-cluster X-Small warehouse forms the basic building block for standard warehouses and has eight processing units; in other words, an X-Small warehouse can handle eight concurrent processes.

Every time you increase the size, you double the number of clusters and, consequently, double the number of processing units.

The number of processing units dictates the degree of parallelism supported by the warehouse. Recalling the earlier discussion from Chapter 2, the query optimizer may parallelize certain operations within a single SQL statement; therefore, you cannot determine exactly how many processing units are either available or awaiting workload at any given time. Later within this chapter, you will see how to measure the number of working concurrent processes within a warehouse and how to measure the number of processes queued awaiting resource availability.

The number of serviced concurrent processes is not the only measure of warehouse capacity. In Chapter 3 I discussed spills to disk and OOMs where lack of memory can be seen as a root cause of some performance issues.

I will discuss how to optimize warehouse workloads later in this chapter.

Warehouse Size and Use Considerations

Every time a warehouse is instantiated, you pay for the first 60 seconds of runtime and thereafter for every second of runtime. The primary objectives must be to do the following:

- Squeeze the maximum performance from every processing unit within a running warehouse

- Reduce warehouse runtime duration

- Deliver maximum end-user performance by optimizing data retrieval

As you might expect, you must test, test, and then test again to inform your warehouse sizing decision. Note that there are many considerations when optimally sizing and using your warehouses.

- Is the workload consistent with historical "steady-state" workloads?

- How many concurrent workloads are running against the warehouse?

- What is your warehouse concurrency set to?

- Are workloads queueing?

- Is warehouse clustering enabled, and if so, to what degree?

- For each workload, are any workloads spilling to disk?

- Is object locking evident?

- Does the warehouse run too frequently?

- Are too many warehouses of same size declared?

- Is there low warehouse cache reuse?

- Is the auto_suspend setting incorrect?

- Is an artificial warehouse size constraint imposed?

- Are the files correctly sized for ingestion?

- Is the warehouse correctly sized for the workload?

- Is serial or parallel logging implemented?

As you can see, a root-cause analysis of performance issues does not automatically lead to increasing the warehouse size; you have several options available.

Warehouse Scaling

In this section; I briefly outline the warehouse scaling options available. Figure 6-1 illustrates how you can scale warehouses.

Figure 6-1. *Warehouse scaling options*

A *cluster* is a segregated grouping of both compute and memory. Warehouses are declared with one or more clusters. At the time of writing, a standard warehouse cluster has eight processing units and 16GB to 24GB memory for which you can nominally assume 20GB memory. As discussed in Chapter 1, in August 2022 Snowflake began to record zero-cost and performance benefits, which accrue due to the periodic replacement of obsolete CSP hardware and optimizer performance releases. You cannot be sure the currently stated eight processing units and nominal 20GB memory will hold true in the future, though any changes will undoubtedly be to the Snowflake customer's benefit.

A single-cluster X-Small warehouse has a single cluster. A 6XL warehouse has 512 processing units and an undisclosed amount of memory. While processing units scale linearly, it is not clear whether memory scales linearly. All clusters allocated to a warehouse share the same allocated memory. You can therefore assume a Small warehouse has 16 processing units and between 32GB and 48GB of memory.

I will discuss the benefits and penalties for the warehouse scaling options later in this chapter considering the opposite perspective: why consolidating warehouses of the same size into fewer warehouses represents a good practice. But I will leave warehouse consolidation until you have a firm understanding of warehouse use.

Scale Up

Scaling up warehouses doubles the number of processing units and increases the amount of memory available to the warehouse. Sizes start with X-Small and a single cluster.

Scaling up does the following:

- Enables query performance improvement

- Reduces spills to disk and OOM errors

- Allows more complex queries

- Doubles the cost of running the warehouse

- Does not auto-scale (manually sized only)

- Adds processing units when enabled through QAS

Here's an example of scaling up by increasing the warehouse size:

```
USE ROLE sysadmin;

ALTER WAREHOUSE ingest_xsmall_wh SET WAREHOUSE_SIZE = 'SMALL';
```

Scale Out

Scaling out retains the same warehouse size while adding clusters up to a maximum of 10 clusters.

Scaling out does the following:

- Increases concurrent processing

- Supports more users

- Increases the cost linearly up to a maximum of 10 clusters

- Auto-scales by adding clusters on demand and then suspending clusters as demand falls

Here's an example of scaling out by adding clusters to a warehouse. Note that additional clusters are instantiated only when the workload requires additional resources.

```
USE ROLE sysadmin;

ALTER WAREHOUSE ingest_xsmall_wh SET
MIN_CLUSTER_COUNT = 1
MAX_CLUSTER_COUNT = 4;
```

In the previous example for the X-Small warehouse, you would expect a single cluster to service between one and eight concurrent requests. When a ninth request arrives, a second cluster is instantiated to service concurrent requests 9 to 16, then a third cluster to service concurrent requests 17 to 24, and finally a fourth cluster to service concurrent requests 25 to 32. As MAX_CLUSTER_COUNT is set to 4, no further clusters will be instantiated.

Scale Across

Scaling across declares more named warehouses of the same size. Scale-across warehouses can also be scaled up and scaled out. The scenarios where you should consider scaling across are less clear than those for scaling up and scaling across.

Scaling across:

- Segments workloads by named warehouse

- Removes resource contention

- Decreases concurrent processing

- Increases consumption costs

Scaling across is achieved by declaring multiple warehouses of the same size.

Figure 6-2 illustrates the number of clusters instantiated and credit consumption per hour alongside benefits for both scaling up and scaling out.

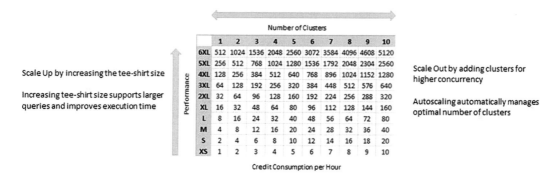

Figure 6-2. *Warehouse scaling options*

Within large complex organizations, you might choose to scale across to explicitly identify each business units resource consumption or to deliberately segregate workload for other reasons. In general, this approach leads to inefficient use of warehouses, and later I explain how query tags can achieve the same result with consolidated warehouses.

Query History

In this chapter, you will make extensive use of `snowflake.account_usage.query_history`, which has latency of up to 45 minutes. You can find more information at `https://docs.snowflake.com/en/sql-reference/account-usage/query_history`.

More immediate monitoring for an individual database can be achieved by using the `query_history` family of table functions for which further information can be found at `https://docs.snowflake.com/en/sql-reference/functions/query_history`.

Throughout this chapter I will use `snowflake.account_usage.query_history` in examples and leave adaptations of `information_schema` for your further investigation.

Background Processes

This investigation will reveal background processing occurs that may be optimizer parallel processing or other non-user-invoked processes. As an example, the statement `show terse SCHEMAS in DATABASE IDENTIFIER('"TPC"') limit 10000` was found when examining the `snowflake.account_usage.query_history` content, and there are many other SQL statements. Background process statements appear to consume trivial resources and run for fractions of a second. The following are the important points to note:

- You cannot control background processes.

- Background processes may run under your application-declared warehouses.

- Background processes appear to resolve internal dependencies.

Should you be concerned about the impact of background process invocation and resource consumption? My opinion is to not be concerned. From my brief investigation, it appears that the background processes correlate to user-invoked SQL statements and therefore are necessary. I would be concerned if the background processes were invoking application-declared warehouses as separate activity unrelated to user-invoked SQL statements, but this does not appear to be the case.

The point is: be aware if you may see SQL statements within `snowflake.account_usage.query_history` that have not been user invoked.

Query Tags

Query tags are distinctly different from object tags and should not be confused with each other.

- Object tags are assigned to declared structural objects.

- Query tags are session-level parameters.

Query tags can be set before a SQL statement is issued, and unset, or set to a new value for subsequent SQL statements. In this manner, you can set individual query tags for every SQL operation within your system.

The ability to query your query history by query tag provides a very fine grain of traceability back to the source when investigating performance issues.

You set a query tag for the connected session as follows:

```
ALTER SESSION SET query_tag = 'Finance';
```

I strongly recommend query tags are implemented to assist later investigations.

An individual query tag can contain up to 2,000 characters and can contain JSON.

```
ALTER SESSION SET query_tag =  '{"Team": "Finance", "Query":
"BusinessLineYTD"}';
```

You can investigate query tag values using the SHOW command:

```
SHOW PARAMETERS LIKE 'query_tag';
```

Then you can extract the "value" programmatically.

```
SELECT "key",
       "value"
FROM    TABLE ( RESULT_SCAN ( last_query_id()));
```

Likewise, you can unset a query tag.

```
ALTER SESSION UNSET query_tag;
```

You can find more information on using query tags at https://docs.snowflake.com/en/sql-reference/parameters#query-tag.

Understanding Workloads

In this section you will investigate where and how both cost and performance improvements can be made, starting by assuming a typical application consumption pattern.

Typical Consumption Pattern

Almost without exception, applications perform three operations, as shown in Figure 6-3.

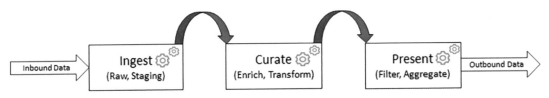

Figure 6-3. *Ingest, curate, present*

Data ingestion consumes compute resource for data pipelines either through an explicitly declared warehouse for the Data Manipulation Language (DML) transaction or through serverless compute such as Snowpipe and connected ELT tooling.

Data curation consumes compute resource through the application of business logic to manipulate and enrich data in the creation of data products in preparation for sale, adding value to organizations.

Data presentation consumes compute resources through filtering, aggregating, and summarizing our data product by customers from marketplace offerings.

Many third-party vendors offer to identify costs savings by monitoring and resizing warehouses according to expected patterns. But there is a pitfall in adopting this approach: where third-party remuneration is predicated upon a percentage of cost savings made by their performance tuning, what happens when the application owner conducts performance tuning outside of the third-party service provided?

My view is that performance tuning must be baked into applications from the outset. By right-sizing your warehouses from the outset and by implementing techniques outlined within this book, you can expect to remove the need to engage third-party tuning tooling.

Default Warehouse Sizing

One of the earliest challenges when developing Snowflake applications is to predict consumption cost against an expected workload pattern. If you are an experienced Snowflake developer, then you know how hard this seemingly simple task is to accurately predict. If you are new to Snowflake, then your Snowflake sales engineer will offer some advice; however, predicting consumption in an ever-changing technical landscape is difficult.

Referring to the assumed application pattern and acknowledging John Ryan who inspired this section (`https://www.analytics.today/`), see Figure 6-4, where I overlaid the nominal warehouse sizes for each operation within an application as a starting point for this discussion.

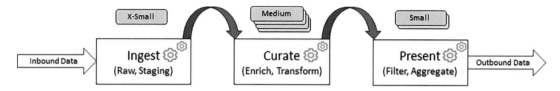

Figure 6-4. *Standard workload warehouse sizing*

John makes the point that each step within an application requires a different default warehouse size. In other words, adopting a single warehouse size, or artificially limiting warehouses to X-Small or Small, can often negatively impact system performance and costs. With the proposed warehouse sizing in Figure 6-4, you at least have a starting point from which to begin your investigations.

Segregating Workload

Snowflake allows you to physically segregate your workloads by declaring warehouses according to logical naming convention. For example, you may declare warehouses by consuming business units such as risk, finance, human resources, etc.

The logical segregation of compute has advantages.

- Rapid identification of consumption at a fine grain according to named warehouses

- Minimal resource contention

While superficially attractive from the perspective of enabling consumption chargeback to the consuming business unit, this approach is inefficient due to warehouses not running fully loaded. Your objective is to maximize throughput while minimizing cost; keeping the load profile high, even to the point of tolerating some queueing, is optimal.

Excessive segregation of workload leads to the following:

- Low utilization resulting in wasted compute

- Reduced ability to benefit from cached data sets

For connected tooling consuming data from the presentation layer, many tools hard-code a single warehouse name within the connection string without the ability to dynamically select an alternative warehouse. As Figure 6-5 suggests, adopting a business unit named warehouse approach will most likely result in resource wastage, even for organizations operating globally 24/7. I prefer fewer named warehouses in general.

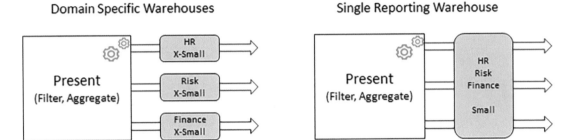

Figure 6-5. *Domain-specific warehouses versus single reporting warehouse*

I suggest replacing business unit–named warehouses in favor of creating a small number of fixed warehouses named according to the purpose that they support: ingestion, curation, and presentation. In multitenant environments, warehouse naming may require the tenant to be prefixed for identification.

Moving from a business domain–based warehouse model to a single reporting warehouse model removes the ability to view consumption by business domain. There is an alternative that requires discipline to both implement and maintain. Alter each session to set a query tag, which persists for every SQL statement issued within the session as discussed earlier within this chapter.

Size Matters

I have mentioned warehouse size several times within this chapter and now will explain "why" size matters.

In a cloud environment, you should not be concerned with the physical server characteristics upon which Snowflake runs, except over time you should expect the CSP hardware to be continually upgraded resulting in steadily improving performance.

The expectation remains that an X-Small warehouse will continue to provide eight processing units and a decent amount of memory to service your queries, all at a reasonable cost per hour of runtime according to Snowflake edition, less any discounts applied.

As you scale up in size, the number of processing units doubles along with more memory allocated. You cannot be sure of the exact memory allocated, but we believe 16GB to 24GB is a reasonable estimate for an X-Small warehouse.

Warehouse performance scalability is linear until the query is no longer resource bound. Figure 6-6 illustrates cost versus performance and shows the "sweet spot" where a nominal query performance cost remains constant and performance significantly improves as warehouse size increases until costs start to rise for little further performance benefit.

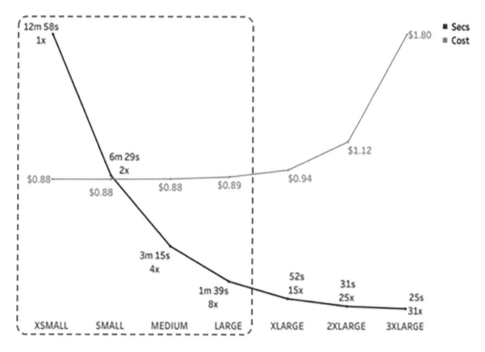

Figure 6-6. *Warehouse size, cost, and performance*

You should not be frightened to use a larger warehouse. While Snowflake charges according to consumption, the stated aim is to drive "good" consumption, that is, warehouse consumption delivering tangible business benefit. Increasing the warehouse size is a single option among, many and you should investigate all options before increasing the warehouse size. One size does not fit all use cases, and you must consider cost implications too.

Dynamic Resizing of Warehouses

Adopting standard warehouse sizing for ingestion, curation, and presentation as shown in Figure 6-5 provides a baseline to consider how you might derive optimal sizing for each part of the application process.

Increasing the warehouse size to improve performance is a viable technique, and some performance tuning applications dynamically determine warehouse size according to stored metrics from previous query runs. Dynamically resizing warehouses can lead to highly inefficient consumption for these reasons:

- If the session disconnects, the warehouse may remain declared larger than expected.

- Resizing warehouses leads to inconsistent runtime costs.

- There is no guarantee the underlying data has not changed; therefore, the resized warehouse may be incorrectly sized.

- Subsequent sessions may use an incorrectly sized warehouse.

- Resizing warehouses flushes the warehouse cache.

We prefer fixed-size warehouses where the warehouse declaration remains constant.

We do not advocate dynamically resizing warehouses and suggest this is a poor approach to performance tuning.

An alternative approach to dynamic resizing of warehouses is to implement QAS, which is discussed later in this chapter.

Tuning the Design

Designing applications for Snowflake under a consumption-based model is distinctly different from designing an application under a traditional on-prem provision-based model. Under an on-prem provision-based model you, do not consider cost as an ongoing factor. The sunk costs for hardware are an up-front initial project cost, and typically you can tune the application according to the constraints of the provisioned environment. Under a consumption-based model, not only must you make optimal use of available processing and storage features, you must also consider the associated costs of using the available processing and storage features.

You must tune your design according to capabilities and pitfalls that the Snowflake consumption-based processing and storage features offer.

Incorrectly setting the warehouse size and runtime frequency is guaranteed to burn credits for no business benefit.

So far within this book you have considered a single Snowflake account, and I have made recommendations for performance tuning individual components. You must also be mindful of how Snowflake operates in the global context of both internal data distribution and within the wider context of distributing data via Snowflake Marketplace, Private Listings, and Secure Direct Data Shares. While I discuss this in depth in a later chapter, the decisions you make when curating data before distribution will have a material impact in terms of timeliness and cost to replicate data.

When porting code from legacy RDBMS to Snowflake, it is possible to have introduced serial processing where common tables are used across two or more processes. An example of serializing parallel processing is when logging information into a Snowflake table. You can address this issue while setting up event logging later within this chapter.

Tuning your design involves processing the minimal subset of data required, in an optimal timeframe, to achieve your business objectives. You must find ways to minimize UPDATE, DELETE, and MERGE operations and prefer INSERT operations instead. This point is relevant insofar as poor performance is visible through warehouse tooling and runtimes, but the underlying database design and DML impact are largely hidden.

Before investigating how to identify optimal warehouse size and use considerations, you must configure the environment, the subject of the next section.

Serial or Parallel Logging

To retain a consistent view of every step throughout the data ingestion, curation, and production of consumption data, processes typically log information into a logging table. In legacy relational database management systems (RDBMSs), you will typically implement a central logging table. Logging will rely upon the online transactional processing (OLTP) capabilities of the SQL engine to handle concurrency where multiple processes attempt to log information at the same time.

Implementing a single table serializes logging operations due to the immutable micro-partition approach. Every DML operation results in new micro-partitions, and therefore high-velocity, low-volume DML operations cause queueing, locking, and inevitable session failure as the transaction workload cannot be fulfilled in a timely manner.

The root cause of process failure during logging information is not directly caused by our warehouse. Resizing our warehouse may help alleviate the symptom of process failure but is an unsafe and potentially costly way to proceed.

We need an alternative approach, and fortunately Snowflake has provisioned event tables to record event information; see for which further information at `https://docs.snowflake.com/en/sql-reference/sql/create-event-table`.

Event Table Implementation

In this section you will create an event table and template code for integrating event logging within your application. Snowflake allows the creation of a single event table per account. Therefore, all events are visible to appropriately entitled roles. In a multi-tenant environment, this may not be desirable as at the time of writing. The segregation of event logging is not possible, and additional steps must be taken to protect logged event information, something I will leave for your further investigation.

I will use Python within this section and assume you have enabled Python (you can learn more about installation and maintenance of tooling in the appendix). You must wait a few minutes until automated provisioning has completed before proceeding.

In this example, create an event table owned by `ACCOUNTADMIN`, grant entitlement to individual roles, and provide a wrapper stored procedure to create events. The supplied code can be rerun in its entirety and is intended to be a template for customization.

Declare identifiers for later use.

```
SET tpc_owner_role     = 'tpc_owner_role';
SET tpc_warehouse_XS   = 'tpc_wh_xsmall';
SET tpc_database       = 'tpc';
SET tpc_public_schema  = 'tpc.public';
```

Set the execution context.

```
USE ROLE      accountadmin;
USE DATABASE  IDENTIFIER ( $tpc_database       );
USE SCHEMA    IDENTIFIER ( $tpc_public_schema );
USE WAREHOUSE IDENTIFIER ( $tpc_warehouse_xs  );
```

Create an event table with a data retention time of seven days; you may want to implement a task to periodically capture logged events before expiry.

```
CREATE OR REPLACE EVENT TABLE tpc.public.monitor_event COPY GRANTS
data_retention_time_in_days = 7;
```

Create one event table for the account.

```
ALTER ACCOUNT SET event_table = 'tpc.public.monitor_event';
```

Set log_level for the desired database.

```
ALTER DATABASE IDENTIFIER ( $tpc_database ) SET log_level = WARN;
```

In this example, you can set LOG_LEVEL at the database level; other object types can also have LOG_LEVEL defined. Setting the LOG_LEVEL parameter to WARN captures all WARN, ERROR, and FATAL messages in the event table. You can find more information on setting log_level at https://docs.snowflake.com/en/developer-guide/logging-tracing/logging-log-level.

Create a Python event logging stored procedure; note that your version of Python may differ from 3.8 shown here:

```
CREATE OR REPLACE PROCEDURE tpc.public.event_logger( P_MESSAGE STRING )
RETURNS VARCHAR
LANGUAGE PYTHON
PACKAGES      = ('snowflake-snowpark-python', 'snowflake-
telemetry-python')
RUNTIME_VERSION = 3.8
HANDLER       = 'run'
EXECUTE AS CALLER
AS
$$
import logging
```

```
logger = logging.getLogger("event_logger")
from snowflake import telemetry

def run ( session, P_MESSAGE ):
   telemetry.set_span_attribute ( "message.proc.run", "begin" )
   telemetry.add_event ( "event_logger", { "message.message_text":
   P_MESSAGE })
   return "SUCCESS"
$$;
```

You can find more information on Python log levels at https://docs.python.org/3/library/logging.html#levels.

Entitle roles to set TRACE LEVEL and LOG LEVEL.

```
GRANT MODIFY         TRACE LEVEL ON ACCOUNT TO ROLE tpc_owner_role;
GRANT MODIFY SESSION TRACE LEVEL ON ACCOUNT TO ROLE tpc_owner_role;
GRANT MODIFY         LOG   LEVEL ON ACCOUNT TO ROLE tpc_owner_role;
GRANT MODIFY SESSION LOG   LEVEL ON ACCOUNT TO ROLE tpc_owner_role;
```

Entitle roles to use event logging.

```
USE ROLE securityadmin;

GRANT USAGE  ON SCHEMA IDENTIFIER ( $tpc_public_schema ) TO ROLE
tpc_owner_role;

GRANT USAGE  ON PROCEDURE tpc.public.event_logger (STRING) TO ROLE
tpc_owner_role;
GRANT SELECT ON TABLE     tpc.public.monitor_event        TO ROLE
tpc_owner_role;
```

Switch to the owner role and set the execution context.

```
USE ROLE      IDENTIFIER ( $tpc_owner_role   );
USE DATABASE  IDENTIFIER ( $tpc_database     );
USE SCHEMA    IDENTIFIER ( $tpc_owner_schema );
USE WAREHOUSE IDENTIFIER ( $tpc_warehouse_xs );
```

Set the session trace_level.

```
ALTER SESSION SET trace_level = ON_EVENT;
```

Create a test event.

```
CALL tpc.public.event_logger ( 'Test event logging' );
```

After calling the test harness to insert an event, you must wait a few minutes before the event is registered before attempting to query the recorded event.

```
SELECT  timestamp                                           AS event_timestamp,
        RESOURCE_ATTRIBUTES [ 'snow.executable.name' ] AS event_source,
        RECORD [ 'name' ]                                   AS event_name,
        RECORD_ATTRIBUTES                                   AS attributes
FROM    tpc.public.monitor_event;
```

You should see the events recorded as shown in Figure 6-7.

EVENT_TIMESTAMP	EVENT_SOURCE	EVENT_NAME	ATTRIBUTES
2023-10-30 12:44:51.292	"EVENT_LOGGER(P_MESSAGE VARCI	"snow.auto_instrumented"	{ "message.proc.run": "begin" }
2023-10-30 12:44:51.290	"EVENT_LOGGER(P_MESSAGE VARCI	"event_logger"	{ "message.message_text": "Test event logging" }

Figure 6-7. *Stored event result set*

Event Logging Integration

You anticipate that event logging will integrate with existing logging procedures by doing either of the following:

- Creating a wrapper stored procedure with an additional parameter directing logging to the original logging stored procedure or to the new Python event based stored procedure.

- Amending the original logging procedure to redirect the logged information into the event table.

In my other book *Maturing the Snowflake Data Cloud*, I demonstrated how to send emails using both JavaScript and Python. Combining events and alerts extends the detection, monitoring, and system management capability further.

Troubleshooting

While configuring event tables, I found these commands useful when troubleshooting:

```
SHOW PARAMETERS LIKE 'event_table' IN ACCOUNT;

SHOW EVENT TABLES;

SHOW EVENT TABLES LIKE '%monitor_event%' IN tpc.public;
```

Workload Predictability

Predictable workloads are key to ensuring declared warehouse sizes remain appropriate for executed queries. You can protect your data ingestion processing by implementing effective monitoring of inbound data volumes and applying tolerance thresholds. Where inbound data sets exceed preset thresholds, suspending processing with associated alerting provides a degree of insulation against either process failure or excessive consumption.

Earlier within this chapter as suggested by John Ryan (`https://www.analytics. today/`), I proposed setting the default warehouse sizes according to their purpose. While this advice remains valid, there are always exceptions for one-off data loads or periodic uploads where data volumes are unpredictable. During our development and delivery process, you must set up the test cases to be representative of expected "steady-state" production workloads. In more complex environments, the expected number of concurrent processes must also be determined to either segregate or scale the workloads.

Workload Monitoring

Optimizing warehouse utilization implies the ability to identify warehouse loads throughout the application daily processing cycle. The application design will be predicated upon handing ingestion, curation, and consumption, and for each, there will be known peaks and troughs in activity. Workloads are readily identified from historical query runtimes and should be reflected into system documentation for reference. In more comprehensive support environments, workload threshold breach detection will provide alerts as events are detected for near real-time investigation. Senior management wants predictability and consistency; no one likes surprises!

Effective monitoring can be performed at the following levels:

- **Individual statement:** Sets the `query_tag` at the session level prior to a statement run

- **Warehouse:** Monitors consumption on a periodic basis

In this section, you will investigate warehouse consumption.

Let's set our monitoring criteria. What is the hourly workload per warehouse?

To begin answering the monitoring criteria, let's create a view `v_warehouse_workload_by_hour` to pre-filter actual work done and summarize the date and hour of execution.

```
CREATE OR REPLACE VIEW v_warehouse_workload_by_hour COPY GRANTS
AS
SELECT warehouse_name,
       start_time,
       end_time,
       query_id,
       query_text,
       total_elapsed_time / 1000  AS total_elapsed_time_in_secs,
       queued_overload_time ,
       transaction_blocked_time,
       DATE_PART (        'YYYY', start_time )||
       LPAD ( DATE_PART ( 'MM',   start_time ), 2, '0' )||
       LPAD ( DATE_PART ( 'DD',   start_time ), 2, '0' )||'_'||
       LPAD ( DATE_PART ( 'HOUR', start_time ), 2, '0' )
                                AS date_time
FROM    snowflake.account_usage.query_history
WHERE   execution_time <> 0
ORDER BY warehouse_name,
        start_time DESC;
```

The view `v_warehouse_workload_by_hour` segments queries by the hour in which they start. The view does not track processes, which start in one hour and continue running into the next hour. This approach is to gain a high-level view of activity; I will leave the refinement of this approach to your further investigation.

The inclusion of `queued_overload_time` facilitates queueing investigation, and the inclusion of `transaction_blocked_time` facilitates object locking investigation; both are addressed shortly.

By using the view `v_warehouse_workload_by_hour`, you can gain insight into how your warehouses are used. For each warehouse you want to know the following:

- Number of SQL statements executed

- Total query execution time

- The date and hour for each group of SQL statements

The next query answers these questions:

```
SELECT warehouse_name,
       COUNT ( query_id )                  AS queries_per_hour,
       SUM ( total_elapsed_time_in_secs ) AS sum_elapsed_secs,
       date_time
FROM   v_warehouse_workload_by_hour
GROUP BY warehouse_name,
         date_time
ORDER BY warehouse_name,
         sum_elapsed_secs DESC,
         date_time       DESC;
```

Let's look at sample output, as shown in Figure 6-8.

WAREHOUSE_NAME	QUERIES_PER_HOUR	SUM_ELAPSED_SECS	DATE_TIME
TPC_WH_XSMALL	20	6.272000	20231029_04
TPC_WH_XSMALL	14	5.584000	20231029_06
TPC_WH_XSMALL	49	2.282000	20231020_01
TPC_WH_XLARGE	18	516.084000	20231020_12
TPC_WH_XLARGE	26	284.058000	20231020_01
TPC_WH_XLARGE	89	186.226000	20231021_02
TPC_WH_XLARGE	12	174.047000	20231021_05

Figure 6-8. *Sample warehouse utilization*

Here you will use the warehouse TPC_WH_XLARGE and a date_time value of
20231021_02 to investigate concurrent process execution.

Before you attempt to derive information for further investigation, you must
understand the result set within its context.

- queries_per_hour is the sum of all queries that begin execution
  within the date_time hour.

- sum_elapsed_seconds is the total for all queries executed within
  the hour.

- date_time shows the workload peak frequency for each warehouse.

- Warehouse billing is per minute on instantiation and per second after
  the first minute has elapsed.

Remember the objectives: to maximize concurrency and to minimize cost. To
achieve the objectives, you must identify how many queries execute concurrently within
the SAME warehouse. You have used two values from Figure 6-8, which should be
replaced with your chosen values before execution.

```
SELECT start_time,
       end_time,
       date_time,
       query_id,
       total_elapsed_time_in_secs
FROM   v_warehouse_workload_by_hour
WHERE  warehouse_name = 'TPC_WH_XLARGE'
AND    date_time       = '20231021_02'
ORDER BY start_time DESC;
```

For the named warehouse and specific date_time, you might identify a result
set as shown in Figure 6-9. Note the objective is to identify overlapping process
execution times.

START_TIME	END_TIME
2023-10-21 02:49:23.971 -0700	2023-10-21 02:49:24.282 -0700
2023-10-21 02:49:06.355 -0700	2023-10-21 02:49:06.415 -0700
2023-10-21 02:49:03.971 -0700	2023-10-21 02:49:04.930 -0700
2023-10-21 02:47:52.359 -0700	2023-10-21 02:47:52.414 -0700
2023-10-21 02:47:40.585 -0700	2023-10-21 02:47:40.609 -0700
2023-10-21 02:47:40.585 -0700	2023-10-21 02:47:40.620 -0700

Figure 6-9. *Processes start and end times*

In Figure 6-9 you can see that the first four records do not overlap; their start_time values do not overlap with an executing process. The last two records do overlap; they start at exactly the same time.

Having understood how process execution start_time values can overlap, you now extract the overlapping subset of records using the same named warehouse and specific date_time, as shown here:

```
SELECT  v1.start_time,
        v1.end_time,
        v1.query_id,
        v1.total_elapsed_time_in_secs,
        v1.date_time
FROM    v_warehouse_workload_by_hour v1
WHERE   EXISTS
        (
        SELECT 1
        FROM    v_warehouse_workload_by_hour v2
        WHERE   v2.start_time <= v1.end_time
        AND     v2.end_time   >= v1.start_time
        AND     v2.date_time   = v1.date_time
        AND     v2.query_id   != v1.query_id
        )
```

```
AND     v1.warehouse_name = 'TPC_WH_XLARGE'
AND     v1.date_time       = '20231021_02'
ORDER BY v1.start_time DESC;
```

Figure 6-10 identifies overlapping process execution times.

START_TIME	END_TIME
2023-10-21 02:47:40.585 -0700	2023-10-21 02:47:40.620 -0700
2023-10-21 02:47:40.585 -0700	2023-10-21 02:47:40.609 -0700
2023-10-21 02:47:39.412 -0700	2023-10-21 02:47:39.521 -0700
2023-10-21 02:47:39.409 -0700	2023-10-21 02:47:39.529 -0700
2023-10-21 02:47:39.409 -0700	2023-10-21 02:47:39.531 -0700

Figure 6-10. *Overlapping process start and end times*

Within Figure 6-10 you can see two groups that overlap; within the groups, processes execute concurrently.

Automated alerting can be implemented by converting the example queries to use `information_schema` table functions executed using task on a periodic timed basis with result sets stored into a local table. I will leave this for your further investigation.

Workload Queueing

Workload queueing occurs within warehouses and represents the number of concurrent processes awaiting service by the warehouse. For this scenario, the number of clusters is not relevant; you are only investigating overload conditions where the warehouse cannot service demand.

Causes of Queuing

Using a single X-Small cluster, if all eight processing units are servicing requests and one more query is executed against this warehouse, you will experience queueing. You may also experience queueing where less than eight processing units are servicing requests but all other warehouse resources are consumed.

Object locking may also cause queueing. Earlier within this chapter you identified multiple concurrent processes logging into a single table as a root cause for object locking.

Occasionally you might experience queueing due to warehouse provisioning. Until the Snowflake background processes and resources become available, our queries will queue.

On the rare occasion when a warehouse fails, the Snowflake background detection and repair processes will "self-heal" by initializing a replacement warehouse for the failed warehouse. Queries hosted on the failed warehouse will either restart or fail.

In real-world use, you have experienced CSP network issues that manifest as excessively long runtimes. Although rare, it is important to capture any outlier runtimes and raise a support ticket as the root cause may not be with your application but instead caused by CSP infrastructure.

Measuring Queueing

There are three ways to observe queueing:

- Snowsight query history

- Account usage store `query_history` view; maximum latency of 45 minutes

- Database information schema `query_history` or `query_history_by_ {session | user | warehouse} TABLE` function

You will first investigate Snowsight query history by selecting "Queued" from within the Status drop-down, as shown in Figure 6-11.

Query History

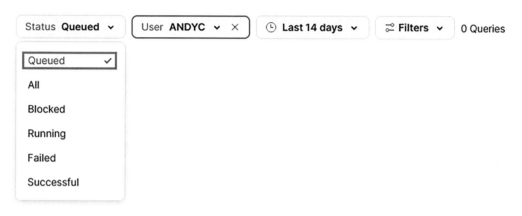

Figure 6-11. *Snowsight query history queueing*

Queueing can be identified using the following SQL statement where queued_ overload_time represents the amount of time waited before execution commences. Any queued_overload_time value greater than zero indicates a query was queued within the corresponding hour.

This query uses information_schema for the current database noting there is no latency but results are retained for only 14 days.

```
SELECT role_name,
       warehouse_name,
       COUNT ( 1 )                      AS num_queued_procs,
       SUM ( total_elapsed_time ) / 1000
                                        AS total_elapsed_time_in_secs,
       DATE_PART (         'YYYY', start_time )||
       LPAD ( DATE_PART ( 'MM',    start_time ), 2, '0' )||
       LPAD ( DATE_PART ( 'DD',    start_time ), 2, '0' )||'_'||
       LPAD ( DATE_PART ( 'HOUR', start_time ), 2, '0' )
                                        AS date_time
FROM   TABLE ( information_schema.query_history())
WHERE  queued_overload_time > 0
```

```
GROUP BY role_name,
         warehouse_name,
         date_time
ORDER BY date_time DESC;
```

This query uses the Account Usage view. Note there is a maximum latency of 45 minutes, and results are retained for a year.

```
SELECT role_name,
       warehouse_name,
       COUNT ( 1 )                          AS num_queued_procs,
       SUM ( total_elapsed_time ) / 1000
                                   AS total_elapsed_time_in_secs,
       DATE_PART (          'YYYY', start_time )||
       LPAD ( DATE_PART ( 'MM',   start_time ), 2, '0' )||
       LPAD ( DATE_PART ( 'DD',   start_time ), 2, '0' )||'_'||
       LPAD ( DATE_PART ( 'HOUR', start_time ), 2, '0' )
                                   AS date_time
FROM   snowflake.account_usage.query_history
WHERE  queued_overload_time > 0
GROUP BY role_name,
         warehouse_name,
         date_time
ORDER BY date_time DESC;
```

The result set exposes the number of processes that were queued within an hour, and for how long, but it does not tell you which queries were queued. However, you now know the role, warehouse, and hour to investigate, and you must check all processes, not just those with queued_overload_time value greater than zero.

```
SELECT query_id,
       query_text,
       total_elapsed_time,
       DATE_PART (          'YYYY', start_time )||
       LPAD ( DATE_PART ( 'MM',   start_time ), 2, '0' )||
       LPAD ( DATE_PART ( 'DD',   start_time ), 2, '0' )||'_'||
       LPAD ( DATE_PART ( 'HOUR', start_time ), 2, '0' )
                                   AS date_time
```

218

```
FROM    snowflake.account_usage.query_history
WHERE   execution_time  > 0
AND     query_type      = 'SELECT'
AND     role_name       = '<YOUR_ROLE_HERE>'
AND     warehouse_name  = '<YOUR_WAREHOUSE_HERE>'
AND     date_time       = '<YYYYMMDD_HH24_HERE>';
```

Scalability will change after each software release for the affected components; the impact should be immediately obvious after the first production run.

You can find more information on queueing at `https://community.snowflake.com/s/article/Understanding-Queuing`.

Resolving Concurrency Issues

In the previous sections, you examined how to identify both warehouse workload and warehouse queueing. This section addresses how to resolve concurrency by offering a number of options.

Regardless of the approach adopted, you must test, test, and then test again to ensure your change has the desired beneficial effect.

Reducing Warehouse Concurrency

The default maximum concurrency level is eight per cluster. Limiting concurrency is applied at the warehouse level and will affect every cluster within the warehouse. Reducing concurrency may be preferrable to ensure more resources are available to support fewer processes:

```
ALTER WAREHOUSE <warehouse_name> SET MAX_CONCURRENCY_LEVEL = 1;
```

You can increase the concurrency level above eight, though you may observe blocking due to insufficient processing resources being available to service demand.

To display the parameter settings for an individual warehouse, use this:

```
SHOW PARAMETERS LIKE 'MAX_CONCURRENCY_LEVEL' IN WAREHOUSE <warehouse_name>;
```

Caveats apply when reducing concurrency; notably, queueing may increase. Therefore, rigorous testing before release into production environments is imperative.

You can find more information on reducing concurrency at `https://docs.snowflake.com/en/user-guide/performance-query-warehouse-max-concurrency`.

Using Summaries, Aggregates, Filters

Examining query profiles often reveals opportunity to implement intermediate objects. The use of materialized views and dynamic tables (currently in public preview) to summarize, aggregate, and subset data often reduces query performance times leading to improved concurrency.

You must be mindful of the additional storage, serverless compute resource, and consequent cost required to support both materialized views and dynamic tables. In high-volume transaction environments, you may see that serverless compute cannot update dependent objects in a timely manner.

Re-timing Processes

Snowflake as a data warehouse is not well suited to high-frequency, low-velocity change. Tuning our design involves questioning the frequency at which you run your ingestion and curation processes. You should identify whether a less frequent ingestion and curation cadence can be tolerated within your service-level agreements (SLAs) to your customers.

Auto-Suspend Setting

As a general principle, you should set your warehouse `AUTO_SUSPEND` setting to be 60 corresponding to one minute. In the vast majority of cases this is sound advice; however, there are a few scenarios where suspending warehouses may not be desirable.

- The warehouse is in continual use. I advocate consolidating warehouses of the same size into a single declaration and using query tags with JSON to differentiate usage as discussed shortly.

- Minimal warehouse lag is desired. Suspending and resuming warehouses incurs a small time penalty.

- A larger value may be beneficial for cache reuse, like when using BI tools.

I recommend developing a thorough understanding of your application profile and warehouse usage before deciding to remove auto suspension.

Snowpipe File Size

The recommended file size for Snowpipe is 100MB to 250MB compressed. Ingesting smaller files generally leads to both increased cost and longer warehouse runtimes. If using serverless compute, you may find your serverless costs higher than expected. Additional costs accrue due to file handling charges, as explained here: `https://www.` `snowflake.com/legal-files/CreditConsumptionTable.pdf`.

Wherever possible, I suggest files are concatenated before load to reduce load costs and runtimes. Note that you may need to strip header and footer records as appropriate.

You can find more information on file sizing at `https://docs.snowflake.com/en/` `user-guide/data-load-considerations-prepare#file-sizing-best-practices-and-` `limitations`.

Artificial Warehouse Size Constraint

In a misguided attempt to control costs, some organizations impose artificial warehouses size limits in an attempt to constrain costs. Such constraints can cause the following:

- Spills to disk

- Reduced concurrency

From our earlier investigation within this book for a nominal query, I identified the optimal warehouse size delivering high performance at a minimal cost. My experience suggests testing and evidence are better indicators of cost and performance, rather than the imposition of artificial constraints.

Object Locking

Object locking is not a warehouse-specific issue except in the sense of many processes attempting to concurrently access objects that create contention, effectively serializing applications. You may see object locking across several warehouses accessing objects concurrently as is the case when logging information.

I addressed using event logging instead of table locking earlier, but remember that this is not the only scenario where serializing of processes may occur.

Object locking is temporal; that is, the same two application components interacting with different data sets may not produce a locking issue. Object locking is caused by two processes attempting to concurrently access a single object. The earliest SQL statement typically acquires an object lock and must complete its operation before releasing the lock to allow the second process to acquire an object lock.

Although the symptoms of object locking appear to be warehouse related and impact process execution time, the root cause is serialized access to a single object for concurrent processes.

This query identifies transactions that experienced locking:

```
SELECT start_time,
       end_time,
       date_time,
       query_id,
       query_text,
       transaction_blocked_time
FROM   v_warehouse_workload_by_hour
WHERE  transaction_blocked_time > 0
ORDER BY start_time DESC;
```

A thorough investigation of transactions and locks is beyond the scope of this book; having introduced the subject I will leave this topic for your investigation. You can find more information at https://docs.snowflake.com/en/sql-reference/sql/show-locks and https://docs.snowflake.com/en/sql-reference/sql/show-transactions.

Consolidating Workloads

Having considered how to identify warehouse performance issues and remediation, you will now consider why consolidating warehouses of the same size into fewer warehouses represents good practice. There are pros and cons to consolidating warehouses of the same size into fewer declarations. The overriding reason is to maximize concurrency while reducing credit consumption. You should aim to run as few warehouses as possible, while utilizing all available processing units for a running warehouse, even to the point of incurring a small amount of queueing.

Didn't I say warehouse queueing is "bad"? In general, overloaded warehouses is a "bad" thing. But running warehouses to their maximum load capability with *occasional and brief* queueing indicates your warehouses are optimally loaded. One size does not fit all, and you cannot always determine warehouse loading particularly where ad hoc queries occur, so a small degree of queueing can usually be tolerated.

Segregating warehouse use by consuming business area can be achieved by implementing query tagging for each discrete process.

As with everything performance related, test, test, and test again.

Load Testing

Load testing encompasses several dimensions. I will not cover a wide range of load testing themes but instead focus on these:

- Snowflake and CSP improvements

- Performance evaluation

- Scalability

- Resource utilization

Although you might imagine that preproduction testing will mimic real-world post-production, in most cases the reality differs. Your pre-production load testing will be indicative at best and serve as a benchmark for future changes. Your pre-production environment must remain constant or your baseline must be re-established. In other words, you are looking for your changes to deliver incremental performance improvements where the underlying platform remains static.

Snowflake and CSP Improvements

In Chapter 1 I made the following points worth reiterating here:

- The optimizer performance has steadily been enhanced over time realizing tangible benefit to overall query execution times.

- CSP hardware replacement programs for obsolete or end-of-life hardware utilize the latest hardware automatically providing performance uplifts.

Although you are not party to the timing of performance optimization or CSP hardware upgrades delivering increased performance, you must be mindful these occur and take them into consideration when conducting load testing.

Organizations adopting a multi-cloud strategy should also be aware that the different hardware implemented across AWS, Azure, and GCP can lead to performance differences for the same feature. Snowflake works hard to provide a consistent experience across all CSPs though performance can vary over time.

Performance Evaluation

To measure improvement, you must establish individual query history for selected SQL statements. Using the captured information, you can compare current performance and project probable future performance over time.

Ideally, you would regression test every release candidate against a static database configuration prior to release to determine whether changes both improve performance and reduce costs.

This metric captures historical information to establish performance profiles for sample queries. The following code illustrates attributes of interest when evaluating performance (note that the Account Usage view has a maximum of 45 minutes latency):

```
SELECT database_name||'.'||schema_name  AS context,
       warehouse_name,
       role_name,
       query_text,
       query_tag,
       compilation_time,
       execution_time,
       total_elapsed_time
FROM   snowflake.account_usage.query_history
WHERE  query_type         = 'SELECT'
AND    warehouse_name IS NOT NULL
AND    execution_status   = 'SUCCESS'
AND    bytes_scanned      > 0
AND    total_elapsed_time > 1000
ORDER BY total_elapsed_time DESC;
```

Where the query tags have previously been set, you can also search by these:

```
SELECT database_name||'.'||schema_name  AS context,
       warehouse_name,
       role_name,
       query_text,
       query_tag,
       compilation_time,
       execution_time,
       total_elapsed_time
FROM   snowflake.account_usage.query_history
WHERE  query_tag            = '<YOUR_GUERY_TAG_HERE>';
```

Metrics capture should be automated with results stored in a local table to prevent aging out after the one-year retention period. I leave this to you for your further investigation.

Parallel Loading

Parallel load scalability is a predictive metric using historical data to provide an indication of how well your configuration will cope with increases in both data volume and concurrent processing. If implemented well and under typical usage conditions, you may be able to predict when problems may occur.

Scalability will be positively impacted by both Snowflake optimizer improvements, CSP hardware improvements, and application deliveries. Scalability will be negatively impacted by additional loads applied to existing warehouses.

Implementing scalability tests requires the creation of a parallel SQL execution engine where multiple query instances are run concurrently. There are several ways to implement a parallel execution engine including the following:

- Externally scheduled jobs via a third-party orchestration tool such as Azure Data Factory, Coalesce, Control M, Autosys, etc.

- Command-line invoked parallel execution engine: Python or other executable.

- Internally invoked parallel execution engine: tasks, stored procedures.

We do not recommend any particular approach, but anticipate the determining factors will be ease of creation, deployment, configuration, and use.

Snowflake-Supplied Sample Load Test

Please be mindful of the difference between multithreading and multiprocesses.

- Multithreading time slices within a single processing core.

- Multitasking uses separate processes and therefore requires more than one processing core.

The next stored procedure uses Snowflake supplied code found at `https://docs.snowflake.com/en/developer-guide/stored-procedure/stored-procedures-python#running-concurrent-tasks-with-worker-processes`.

The sample stored procedure does not invoke multiple processes.

```
CREATE OR REPLACE PROCEDURE joblib_multiprocessing_proc(i INT)
  RETURNS STRING
  LANGUAGE PYTHON
  RUNTIME_VERSION = 3.8
  HANDLER = 'joblib_multiprocessing'
  PACKAGES = ('snowflake-snowpark-python', 'joblib')
AS $$
import joblib
from math import sqrt

def joblib_multiprocessing(session, i):
  result = joblib.Parallel(n_jobs=-1)(joblib.delayed(sqrt)(i ** 2) for i in
  range(10))
  return str(result)
$$;
```

Now invoke the sample stored procedure:

```
CALL joblib_multiprocessing_proc ( 8 );
```

Figure 6-12 shows the query profile and query history for a single invocation of the sample stored procedure illustrating parallelism is implemented as multithreading, not multiprocessing.

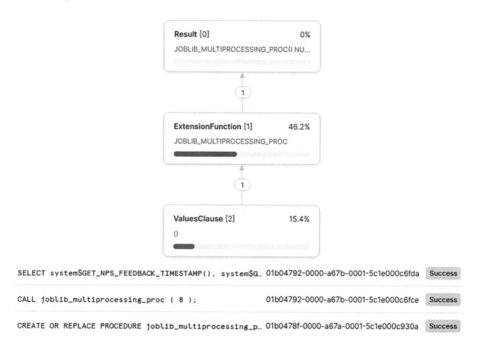

Figure 6-12. *Stored procedure query profile*

You can find more information on joblib at `https://joblib.readthedocs.io/ en/latest/parallel.html`. Of specific interest is the section in parallelism found at `https://joblib.readthedocs.io/en/latest/parallel.html#thread-based- parallelism-vs-process-based-parallelism`.

Tasks and Streams

While a load test process may be created, or even better, generated with a pattern-based, code-templated approach to create tasks and streams, we do not advocate using this approach for load testing.

The act of physically creating and deploying tasks and streams for large-scale parallel load testing is excessive. For example, testing X-Small warehouse scaling out requires more than eight tasks and eight streams with associated long-running SQL statements to create an environment where queueing may be observed. Once testing is complete, the code should be removed ready for the next iteration where objects and attributes may have changed.

What you need is something more flexible than what Snowflake supplies "out of the box," an externally parallelized process that does not rely upon Snowflake structures but can be data-driven using metadata tables.

You will do this next.

External Parallelism Explained

Figure 6-13 illustrates a single X-Small cluster warehouse servicing up to eight concurrent processes shown as Process 1 through Process 8.

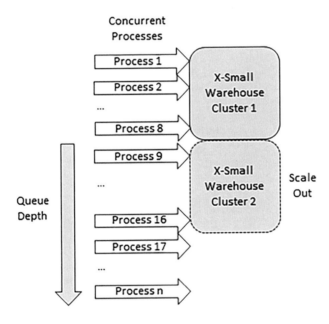

Figure 6-13. *Warehouse concurrent processing explained*

When a new process is invoked against the single X-Small cluster warehouse, the process will be queued as no processing unit is available to service the request. At this point, either you wait for a running process to complete, or we set the X-Small warehouse clustering to scale out and allow the instantiation of a second cluster providing eight more processing units, Process 9 through Process 16. For an X-Small warehouse, you may set the clustering factor to instantiate up to 10 clusters, therefore providing up to 80 concurrent processing units.

Queue depth represents the number of processes waiting for a warehouse processing unit to become available.

Figure 6-14 illustrates how you can monitor clustering from Snowsight though you must first add Cluster Number as a reporting attribute as shown. You have removed some attributes from the displayed image.

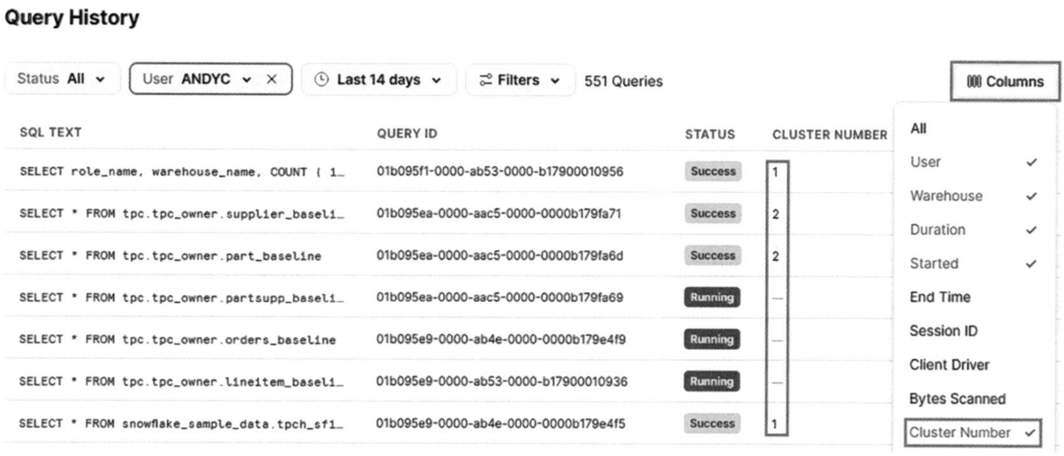

Query History

SQL TEXT	QUERY ID	STATUS	CLUSTER NUMBER	
SELECT role_name, warehouse_name, COUNT (1_	01b095f1-0000-ab53-0000-b17900010956	Success	1	All
SELECT * FROM tpc.tpc_owner.supplier_baseli_	01b095ea-0000-aac5-0000-0000b179fa71	Success	2	User ✓
SELECT * FROM tpc.tpc_owner.part_baseline	01b095ea-0000-aac5-0000-0000b179fa6d	Success	2	Warehouse ✓
SELECT * FROM tpc.tpc_owner.partsupp_baseli_	01b095ea-0000-aac5-0000-0000b179fa69	Running	—	Duration ✓
SELECT * FROM tpc.tpc_owner.orders_baseline	01b095e9-0000-ab4e-0000-0000b179e4f9	Running	—	Started ✓
SELECT * FROM tpc.tpc_owner.lineitem_baseli_	01b095e9-0000-ab53-0000-b17900010936	Running	—	End Time
SELECT * FROM snowflake_sample_data.tpch_sf1_	01b095e9-0000-ab4e-0000-0000b179e4f5	Success	1	Session ID

Figure 6-14. *Warehouse concurrent processing explained*

Running queries do not report their cluster; this attribute is populated after the query completes.

Create an External Parallelism Component

My objective for this section is to create an external parallelism component to implement concurrent processing while demonstrating how to monitor concurrent processing.

With the kind assistance of my colleague and friend Ramya Purushothaman, we offer this Python parallel code as the starting point for your further investigation and extension.

Please refer to the appendix for instructions on installing Python. There are other options available such as Visual Studio (https://code.visualstudio.com/) and Miniconda (https://docs.conda.io/projects/miniconda/en/latest/) though the following instructions reference the appendix installation.

Assuming Python is installed, open a command console and type the following:

```
python --version
```

Figure 6-15 shows the expected response, noting your version may differ.

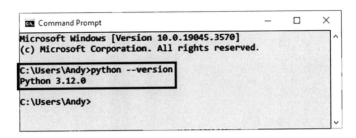

Figure 6-15. *Python version*

Create a new file called `parallel_query.txt` in the same directory as your command window opens up; in this example, it's `C:\Users\Andy`, but your default directory will differ.

Your Snowflake account can be derived from this:

```
SELECT current_account();
```

The account string should look like this, noting your region and CSP may differ:

```
<YOUR_ACCOUNT>.eu-west-2.aws
```

The following query performs these operations:

- Connects to Snowflake

- Creates 10 sample long-running queries

- Executes 10 sample queries

- Periodically reports execution status to console

Modify the code to suit your need noting you must change login credentials.

Copy the following text into `parallel_query.txt`, replacing placeholders with the correct values to connect into your environment:

```
import snowflake.connector
import os
import pandas as pd
import time

def get_status(cur_list,sf_conn):
    print(f'Check the status of the query list : {cur_list}')
```

```python
        status=[]
        df=pd.DataFrame(columns=['Query_id','Status'])
        arr=cur_list
        for query_id in cur_list:
            status_for_the_query=sf_conn.get_query_status(query_id).name
            status.append(status_for_the_query)
            df=df._append({'Query_id' : query_id,'Status':status_for_the_
            query},ignore_index=True)
        if status.count('RUNNING') >1:
            del status[:]
            print(df)
            print('One or more commands still running')
            time.sleep(5)
            get_status(arr,sf_conn)
        else:
            print('All commands execution done!')
            print(df)
        return

def main():
    #Establish snowflake connection
    conn=snowflake.connector.connect(
            user       = '<YOUR USERNAME>',
            password   = '<YOUR PASSWORD>',
            account    = ',YOUR_ACCOUNT>.<REGION>.<CSP>',
            warehouse  = 'tpc_wh_xsmall',
            database   = 'tpc',
            schema     = 'tpc_owner',
            role       = 'tpc_owner_role'
            )
    print('Connected to Snowflake')
    cur=conn.cursor()
    #SQL Statement
    sql1='SELECT * FROM snowflake_sample_data.tpch_sf1000.lineitem'
    sql2='SELECT * FROM snowflake_sample_data.tpch_sf1000.orders'
    sql3='SELECT * FROM snowflake_sample_data.tpch_sf1000.partsupp'
```

```
    sql4='SELECT * FROM snowflake_sample_data.tpch_sf1000.part'
    sql5='SELECT * FROM snowflake_sample_data.tpch_sf1000.supplier'
    sql6='SELECT * FROM tpc.tpc_owner.lineitem_baseline'
    sql7='SELECT * FROM tpc.tpc_owner.orders_baseline'
    sql8='SELECT * FROM tpc.tpc_owner.partsupp_baseline'
    sql9='SELECT * FROM tpc.tpc_owner.part_baseline'
    sql10='SELECT * FROM tpc.tpc_owner.supplier_baseline'
    df=pd.DataFrame(columns=['Query'],data=[[sql1],[sql2],[sql3],[sql4],
    [sql5],[sql6],[sql7],[sql8],[sql9],[sql10]])
    query_ids=[]

    #Iterating through Dataframe and submit query
    for index,row in df.iterrows():
        print(row['Query'])
        cur.execute_async(row['Query'])
        query_id=cur.sfqid
        query_ids.append(query_id)
        print(f'list of query_ids {query_ids}')
    #Getting status of queryid
    get_status(query_ids,conn)
    #Fetching Result of Query
    for query_id in query_ids:
        cur.get_results_from_sfqid(query_id)
        results = cur.fetchall()

if __name__ == '__main__':
    main()
```

Close the file and rename it to parallel_query.py."

The supplied file contains two SQL statements to run in parallel; I will leave the expansion of this script to you.

To build a stand-alone executable, run this:

```
pyinstaller --onefile parallel_query.py
```

Figure 6-16 shows the abbreviated screen output, noting the resultant location of the built executable; in this example, it's C:\Users\Andy\dist\parallel_query.exe.

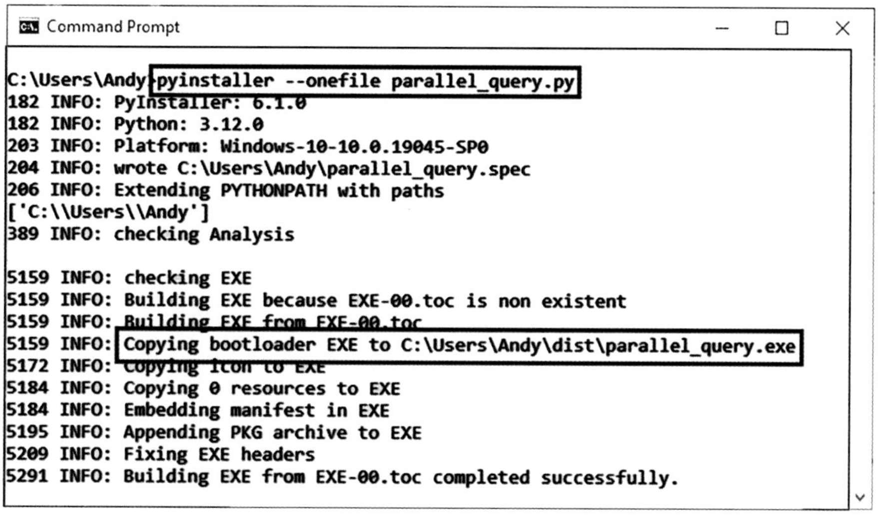

Figure 6-16. *Creating a Python executable*

Testing External Parallelism

With the executable created, you can invoke the executable directly at the command line, noting your path will differ from mine.

```
C:\Users\Andy\dist\parallel_query.exe
```

With our executable running, you may see these errors:

- Failed to import ArrowResult. No Apache Arrow result set format can be used.

- ImportError: No module named 'snowflake.connector.snow_logging'.

Neither error prevents the code from running and, for the purposes of this test, can be ignored.

Figure 6-17 shows the sample output.

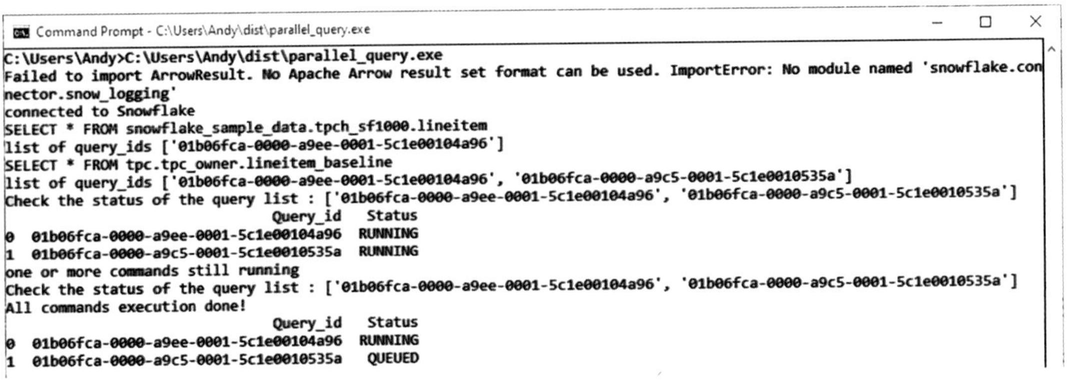

Figure 6-17. *Parallel execution example*

Using the identified query IDs from Figure 6-17, you are able to monitor our warehouse TPC_WH_XSMALL for queueing. For monitoring, you use the table function to remove latency as you require immediate results.

Monitoring Queueing

To observe queueing, you will reuse the monitoring query from earlier.

```
SET tpc_warehouse_S  = 'tpc_wh_xsmall';

USE WAREHOUSE IDENTIFIER ( $tpc_warehouse_xs );

SELECT role_name,
       warehouse_name,
       COUNT ( 1 )                      AS num_queued_procs,
       SUM ( total_elapsed_time ) / 1000
                                        AS total_elapsed_time_in_secs,
       DATE_PART (          'YYYY', start_time )||
       LPAD ( DATE_PART ( 'MM',   start_time ), 2, '0' )||
       LPAD ( DATE_PART ( 'DD',   start_time ), 2, '0' )||'_'||
       LPAD ( DATE_PART ( 'HOUR', start_time ), 2, '0' )
                                        AS date_time
FROM    snowflake.account_usage.query_history
WHERE   queued_overload_time > 0
```

```
GROUP BY role_name,
         warehouse_name,
         date_time
ORDER BY date_time DESC;
```

If the warehouse you are monitoring is overloaded at the point of monitoring and you attempt to use the *same* warehouse to identify monitoring information, you will further overload our warehouse as shown in Figure 6-18.

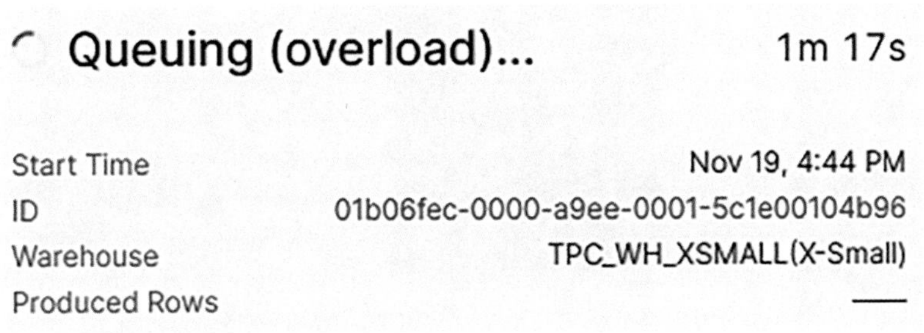

Figure 6-18. *Monitoring warehouse overload condition*

Monitoring *must* use a dedicated warehouse to prevent further overloading of monitored warehouses.

In addition to running a query to monitor queueing, Snowsight offers a graphical representation. Set your role to ACCOUNTADMIN and then using the administrative toolbar navigate to Admin ➤ Warehouses. From the presented list of warehouses, select your desired warehouse, and then select an individual date to expose the pop-up information for the day, as shown in Figure 6-19.

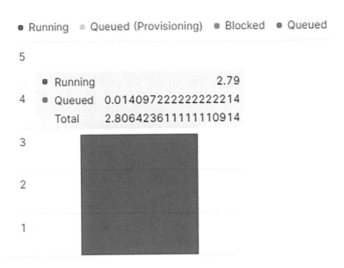

Warehouse Activity

● Running ● Queued (Provisioning) ● Blocked ● Queued

● Running		2.79
● Queued	0.0140972222222222214	
Total	2.8064236111111110914	

Figure 6-19. *Snowsight single warehouse monitoring*

Within the day summary information, you can see most of the warehouse activity is Running with a very small amount of Queued activity. At first glance and recognizing this is a single-day summary over 24 hours, the graphic suggests the warehouse is running optimally as you should tolerate a small amount of queuing to indicate the warehouse is optimally loaded. However, some caveats apply:

- You cannot tell whether warehouse activity is linear throughout the day.

- A single-day warehouse activity may not represent "steady-state" but instead be anomalous.

Snowsight provides filters to select the time period; I will leave experimentation for your further investigation.

You must be careful when extracting information from a very low data sample. Let's now examine a two-week data sample, as shown in Figure 6-20.

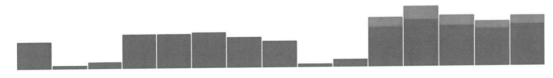

Figure 6-20. *Snowsight two-week warehouse monitoring*

Figure 6-20 tells you that the warehouse is now experiencing queueing. Something changed, but what?

- The processes may take longer to execute because of higher data volumes.

- The data profile may have changed. UPDATE and MERGE operations are very expensive compared to INSERT and DELETE operations.

- A software release may have changed our application execution profile resulting in higher concurrency than expected.

In this particular example, a software release caused excessive queueing.

The important point is to recognize what our monitoring is telling you. With appropriate monitoring, detection, alerting, and visualization tooling, you will be able to quickly identify and rectify issues.

Noting the presence of excessive queueing, scaling out by adding clusters resolved the issues.

Restricting Resource Consumption

So far within this chapter, we have focused on identifying root causes for performance issues from a warehouse perspective. I covered SQL performance tuning, micro-partitions, and cluster key tuning issues in earlier chapters.

Snowflake supplies options for controlling resource consumption, but none addresses performance root causes. However, all limit excessive consumption by preventing uncontrolled query execution. These are as follows:

- STATEMENT_TIMEOUT_IN_SECONDS

- USER_TASK_TIMEOUT_MS

- MAX_CONCURRENCY_LEVEL

- Resource monitors

You can consider these options as instruments of last resort. That is, if all else fails, these options will prevent excessive consumption, but none offers insight to do root-cause analysis.

If you are implementing resource consumption limits, consider the implications to your application where data may be left in an inconsistent state by aborted processes.

Most organizations do not explicitly set these parameters and instead rely upon default settings. Incorrectly setting these values may result in unexpected consequences. You can consider reviewing parameter settings within the context of typical and expected application behavior to be an important step in restricting resource consumption.

I briefly discuss the listed options next.

STATEMENT_TIMEOUT_IN_SECONDS

STATEMENT_TIMEOUT_IN_SECONDS is the maximum time specified in seconds for which an individual SQL statement will run before being cancelled by Snowflake.

This parameter can be set for either session or user or warehouse and defaults to 172,800 seconds (2 days) with a maximum value of 604,800 (7 days).

Where STATEMENT_TIMEOUT_IN_SECONDS for both the session and warehouse is set differently, the lowest nonzero value takes precedence. For example:

- The warehouse timeout set to 1,000 seconds.

- The session timeout set to 500 seconds.

The lowest nonzero value takes precedence, in this example, 500 seconds.

To show the warehouse settings, use this:

```
SHOW PARAMETERS IN WAREHOUSE;
```

And for an individual warehouse, use this:

```
SHOW PARAMETERS IN WAREHOUSE tpc_wh_xsmall;
```

You can find more information at https://docs.snowflake.com/en/sql-reference/parameters#statement-timeout-in-seconds.

STATEMENT_QUEUED_TIMEOUT_IN_SECONDS

STATEMENT_QUEUED_TIMEOUT_IN_SECONDS is the maximum time specified in seconds for which an individual SQL statement will remain *queued* before being cancelled by Snowflake.

This parameter can be set for both session and warehouse and defaults to 0 seconds; there is no maximum value:

- The warehouse timeout is set to 120 seconds.

- The session timeout is set to 60 seconds.

The lowest nonzero value takes precedence, in this example, 60 seconds.

To show the warehouse settings, use this:

```
SHOW PARAMETERS IN WAREHOUSE;
```

For an individual warehouse, use this:

```
SHOW PARAMETERS IN WAREHOUSE tpc_wh_xsmall;
```

You can find more information at https://docs.snowflake.com/en/sql-reference/parameters#statement-queued-timeout-in-seconds.

USER_TASK_TIMEOUT_MS

USER_TASK_TIMEOUT_MS is the maximum time a single run for an individual task will run for before being cancelled by Snowflake.

This parameter defaults to 360000 milliseconds (1 hour) with a maximum value of 86400000 milliseconds (1 day).

STATEMENT_TIMEOUT_IN_SECONDS may take precedence over USER_TASK_TIMEOUT_MS; refer to the documentation for details.

You can find more information at https://docs.snowflake.com/en/sql-reference/parameters#user-task-timeout-ms.

MAX_CONCURRENCY_LEVEL

This was previously discussed within this chapter and is another means to restrict resource consumption.

To show the account setting, use this:

```
SHOW PARAMETERS LIKE 'MAX_CONCURRENCY_LEVEL' IN ACCOUNT;
```

To display the parameter settings for an individual warehouse, use this:

```
SHOW PARAMETERS LIKE 'MAX_CONCURRENCY_LEVEL' IN WAREHOUSE <warehouse_name>;
```

You can find more information at `https://docs.snowflake.com/en/sql-reference/parameters#max-concurrency-level`.

Resource Monitors

Snowflake provides resource monitors as a means to control warehouse consumption. This is a reactive approach to limit costs once the specified threshold has been reached.

A detailed explanation and walk-through of resource monitors is beyond the scope of this chapter but mentioned for completeness.

You can find more information on resource monitors at `https://docs.snowflake.com/en/user-guide/resource-monitors`.

Serverless Compute

Serverless compute enables the seamless provision of compute without the full provision of a warehouse. You can consider serverless compute to be a fractional warehouse where consumption is priced per second, instead of the full cost for the first minute and then per second thereafter.

Snowflake features increasingly offer serverless compute for cost-effective and simple implementations, but the costs can quickly escalate. Some features use serverless compute as a baked-in feature of the core product. You must be mindful of these hidden costs.

Table 6-1 illustrates serverless compute components along with a brief summary of the capability provisioned as derived from `https://docs.snowflake.com/en/user-guide/cost-understanding-compute#serverless-credit-usage`.

Table 6-1. *Serverless Compute Components*

Component	Feature	Compute
Automatic clustering	Automated background maintenance of each clustered table, including initial clustering and reclustering as needed	Serverless only
External tables	Automated refreshing of the external table metadata with the latest set of associated files in the external stage and path	Serverless only
Materialized views	Automated background synchronization of each materialized view with changes in the base table for the view	Serverless only
Query acceleration service	Execution of portions of eligible queries	Serverless only
Replication	Automated copying of data between accounts, including initial replication and maintenance as needed	Serverless only
Search optimization service	Automated background maintenance of the search access paths used by the search optimization service	Serverless only
Snowpipe	Automated processing of file loading requests for each pipe object	Serverless or warehouse
Snowpipe streaming	Automated processing of file loading requests for each pipe object; currently INSERT only	Serverless only
Tasks	Scheduled tasks	Serverless or warehouse

Without addressing every aspect of serverless compute, you will next focus on three aspects where the use of serverless compute requires a little more thought when designing our applications.

Snowpipe

Snowpipe offers built-in capability to rapidly ingest files where serverless compute can be used instead of a dedicated warehouse. The most appropriate compute decision will be informed by the following:

- Source file size; the optimal size is 100MB to 250MB per file

- The number of concurrent loads

Low-volume, high-frequency concurrent Snowpipe loads using serverless compute are highly likely to incur greater cost than comparable loads configured to use a dedicated warehouse.

Please refer to my earlier book *Building the Snowflake Data Cloud* for a Snowpipe working example using a dedicated warehouse.

Tasks

When tasks are declared, you have two options.

- Explicitly define a named warehouse for every task invocation.

- Enable serverless tasks where Snowflake determines the appropriate warehouse size after a few runs.

Snowflake makes every effort to simplify their product. Serverless tasks are an example where reliance upon automated service provision may not deliver optimal costs or performance benefits.

If the question is, "Which is best? Dedicated or serverless compute?" our answer must be "It depends...."

Our view will be informed by the following:

- Predictability of workload size in terms of data set and DML operations

- Whether the workload is low velocity, i.e., infrequent or high velocity

- Understanding how a dedicated warehouse parallelizes concurrent operations

Unlike explicitly declared warehouses, serverless compute is charged by the second but can be expensive if misapplied. There is no single answer for all use cases. Testing under real-world scenarios coupled with proactive monitoring of our production application costs further informs our decision-making, and we are reminded of earlier advice for a typical consumption pattern.

Serverless compute consumed by tasks can be monitored using the Account Usage Store view `serverless_task_history`; you can find further information at `https://docs.snowflake.com/en/sql-reference/account-usage/serverless_task_history`.

Query Acceleration Service

QAS enables the offloading of workload sections to additional shared compute resource, in essence, parallelizing parts of our workload. QAS is intended to prevent over-provisioning of warehouse size by enabling enough additional processing units to efficiently process the workload.

QAS supports these SQL commands:

- SELECT

- INSERT

- CREATE TABLE AS SELECT (CTAS)

You can find more information on QAS-supported commands at https://docs. snowflake.com/en/user-guide/query-acceleration-service#supported-sql-commands.

QAS can be explicitly enabled for each warehouse and scaled as required. Recalling our earlier discussion on warehouse processing unit provision, each cluster has eight processing cores. Enabling QAS enables additional processing units to be added to each cluster within a warehouse, these are spun up as demand increases and shut down when demand falls. As with all serverless compute, service provision is charged by the second.

However, not all workloads are suitable for parallelization, and there are no absolute criteria where QAS is guaranteed to add benefit. The following are the general guidelines for queries eligible for QAS:

- Highly selective filters

- Contains aggregations

- Low cardinality GROUP BY

- Source tables with high number of micro-partitions

- Includes ORDER BY without LIMIT

Of all the listed guidelines, the most important consideration is cardinality of data in determining whether QAS will be used. There is no cost to enabling QAS, and no costs are incurred if QAS is unused.

Snowflake provides an Account Usage Store view called `QUERY_ACCELERATION_ELIGIBLE` to identify queries suitable for QAS. You can find more information at `https://docs.snowflake.com/en/sql-reference/account-usage/query_acceleration_eligible` and `https://docs.snowflake.com/en/sql-reference/functions/system_estimate_query_acceleration`.

As with tasks, if the question is "Should we enable QAS?" the answer must be "It depends upon the nature of your queries."

There is benefit in enabling QAS as part of the consumption processes where you might want additional processing units where unpredictable loads occur and your queries are most likely to conform to the previous list. You will not see any benefits for ingestion processes. Complex logic deployed as part of an ETL/ELT process more typically associated with data curation may benefit from enabling QAS but is counterbalanced by the expectation of having larger warehouses for data curation.

Summary

This chapter began by establishing some foundational information for warehouses and installed tooling.

I then covered workloads by investigating how you might have inadvertently over-provisioned your warehouses resulting in additional costs. Noting the need to identify consumption by business domain, I proposed the use of query tags.

When considering application logging, concurrent processes are often serialized due to the implementation of a single log table. As Snowflake uses immutable micro-partitions, a single log table can cause process queueing or timeouts. I offered a replacement using event tables, a new Snowflake feature proved to significantly reduce resource contention but noted there is a single event table for the whole account.

Several factors affect warehouse performance, and you worked through examples for each. I also offered some mitigating actions and warehouse settings to use. Increasing the warehouse size is not the only option, and a comprehensive understanding of all available options helps when resolving issues.

I discussed load testing extensively and provided guidance on how to create a load test module and identify parallel processing workloads. I then covered queueing and identified options for monitoring queueing.

Snowflake supplies options to restrict runaway resource consumption. Note that these are a blunt instrument and offer little insight into remediating the underlying operations.

Taking action to mitigate warehouse performance is dependent upon identifying the root cause. Increasing warehouse size is not a silver bullet.

The next chapter will cover search optimization.

CHAPTER 7

Search Optimization Service

Earlier chapters discussed clustering, materialized views, dynamic tables, and the Query Acceleration Service (QAS). In this chapter I cover the Search Optimization Service (SOS) as an alternative but complementary approach.

At a fundamental level, query optimization resolves micro-partition pruning to reduce the number of micro-partitions accessed for queries. Viewed from a pure micro-partition pruning perspective, without considering summaries and aggregates, SOS adopts a different approach to enable micro-partition pruning, thereby extending the range of options available to improve query optimizer efficiency.

SOS is a little-understood but very powerful Snowflake service focused on pre-building optimized data structures called *search access paths* maintained via serverless compute. As you will discover in this chapter, search access paths only reference micro-partitions containing explicitly referenced values. For those familiar with legacy RDBMSs, you might consider search access paths to be single-attribute, index-like constructs as they perform similar functionality.

Every serverless compute capability incurs compute cost, and SOS also incurs additional storage costs; therefore, we advise both caution with thorough investigation and testing of appropriate use cases before enabling SOS in your production applications.

Not every query will benefit from SOS; you must align SOS enablement with optimal consumption usage. Figure 7-1 suggests where each serverless compute feature offers optimal benefits; note that SOS is targeted at data consumption from the Presentation layer.

© Andrew Carruthers 2024
A. Carruthers, *Tuning the Snowflake Data Cloud*, https://doi.org/10.1007/979-8-8688-0379-6_7

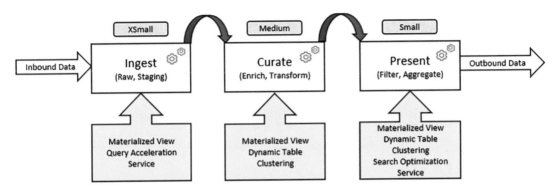

Figure 7-1. *Serverless compute feature use*

Figure 7-1 is illustrative only and suggests where features may be best placed for optimal performance and cost; your implementation may differ. I suggest materialized views may be used for ingest where the typical use case is to flatten JSON and not for other performance reasons. The use case for materialized views in the curation step is to express alternate cluster keys to facilitate a wider range of query predicates. For the Presentation layer, the use case is more closely aligned to performance optimization through summarization, aggregation, and pre-filtering prior to consumption.

SOS can be thought of as an overlay directly related to clustering; if a table is re-clustered, SOS should first be dropped, the table re-clustered, and then SOS enabled. SOS is intended to easily accelerate queries with selective predicates where tables have a high number of micro-partitions. Both statements provide insight into how SOS operates internally. SOS provides alternative mappings for micro-partition lookups to improve query lookup pruning.

SOS is not a silver bullet and must be selectively implemented; we discuss how to make an effective determination for use later in this chapter.

You can find the SOS documentation at `https://docs.snowflake.com/en/user-guide/search-optimization-service`.

Search Optimization Service Explained

Enabling SOS for a table does not imply full coverage of all the possible query predicate operations. By default, an SOS-enabled table creates EQUALITY search access paths only. A wider range of search access paths should be created by deliberate intent on an attribute-by-attribute basis according to known query predicates. I discuss optimal usage scenarios later in this chapter; note that there are many exclusions too.

Search optimization should be applied retrospectively and cannot pro-actively predict future use but must be part of a well-considered holistic approach to application performance tuning. We suggest SOS is enabled in production systems where predictable workloads exist and baseline system performance has been established.

While QAS has no storage component and is focused on compute, QAS can accelerate queries within specific boundaries and may be used with SOS where both services complement each other. You can fine more information on QAS and SOS interaction at https://docs.snowflake.com/en/user-guide/search-optimization-service#compatibility-with-query-acceleration.

SOS implements an alternative to clustering by creating search access paths for each enabled table. Search access paths may take time to create, and any changes to the underlying table content will need to be reflected into the search access paths; note that serverless compute is an asynchronous background process.

You must also be aware of both storage and compute cost implications of enabling SOS. As mentioned, Snowflake provides a suite of services, and you may not need to use every available feature. For example, if a non-SOS enabled query takes two seconds to fulfil, will the same SOS-enabled query fulfilled in one second make enough difference to justify the cost?

Before investigating how to implement SOS, let's examine where SOS can add benefit.

Optimal Use Scenarios

According to SOS's design intent, SOS is targeted at tables with high numbers of micro-partitions for queries with selective predicates.

Where explicitly declared, SOS performs optimally against large tables when returning small subsets of data for highly selective query predicates. Here are some examples:

- Table attribute = `<value>`

- Table attribute `IN ( <value_1>, <value_2>… )`

You can find more information at `https://docs.snowflake.com/en/user-guide/` `search-optimization/point-lookup-queries`.

Partial attribute value and regular expressions can also benefit from SOS.

- Table attribute `LIKE` (and variants)

- Table attribute `REGEXP`

You can find more information at `https://docs.snowflake.com/en/user-guide/` `search-optimization/substring-queries`.

Semi-structured queries can also benefit from SOS. You can find more information at `https://docs.snowflake.com/en/user-guide/search-optimization/semi-` `structured-queries`.

Likewise, geospatial queries can also benefit from SOS. You can find more information at `https://docs.snowflake.com/en/user-guide/search-optimization/` `geospatial-queries`.

Having identified where SOS can benefit queries, you can now examine where SOS cannot be used.

Excluded Use Scenarios

Not every query will benefit from search optimization; notably SOS does not support these scenarios:

- External tables

- Materialized views

- Columns defined with a `COLLATE` clause

- Column concatenation

- Analytical expressions

- Casts on table columns (except for fixed-point numbers cast to strings)

- Floating-point data types

- GEOMETRY data type

You can find more information on excluded scenarios at https://docs.snowflake.com/en/user-guide/search-optimization/queries-that-benefit#queries-that-do-not-benefit-from-search-optimization.

Search Optimization Implementation

Having identified scenarios where SOS can provide performance optimization, let's examine how to implement SOS.

You can start by reusing the previously created TPC environment.

```
SET tpc_owner_role   = 'tpc_owner_role';
SET tpc_warehouse_XS = 'tpc_wh_xsmall';
SET tpc_database     = 'tpc';
SET tpc_owner_schema = 'tpc.tpc_owner';
```

Enabling SOS requires adding the entitlement ADD SEARCH OPTIMIZATION to the role tpc_owner_role.

```
USE ROLE securityadmin;
```

```
GRANT ADD SEARCH OPTIMIZATION ON SCHEMA IDENTIFIER ( $tpc_owner_schema ) TO
ROLE IDENTIFIER ( $tpc_owner_role   );
```

Set the execution context.

```
USE ROLE      IDENTIFIER ( $tpc_owner_role   );
USE DATABASE  IDENTIFIER ( $tpc_database     );
USE SCHEMA    IDENTIFIER ( $tpc_owner_schema );
USE WAREHOUSE IDENTIFIER ( $tpc_warehouse_xs );
```

Estimating Table Search Optimization Costs

As previously discussed, you should selectively enable SOS on a table-by-table basis according to usage and perceived benefits. I will discuss how to enable SOS on an attribute by attribute basis later in this chapter; within this section I am establishing broad principles.

You should note SOS estimates are just that: an SOS estimate does not guarantee real-world delivery of suggested benefit. The Snowflake documentation suggests actual realized costs can vary by 50 percent or more. You can find more information at `https://docs.snowflake.com/en/sql-reference/functions/system_estimate_search_optimization_costs`.

Invoking cost estimation for a table can be done like this:

```
SELECT system$estimate_search_optimization_costs
( 'tpc.tpc_owner.lineitem_baseline' );
```

The returned JSON record when formatted using `https://jsonformatter.org/` and annotated with comments looks like Figure 7-2.

```json
{
  "tableName": "LINEITEM_BASELINE",
  "searchOptimizationEnabled": false,
  "costPositions": [
    {
      "name": "BuildCosts",
      "costs": {
        "value": 10.82735,
        "unit": "Credits"
      },
      "computationMethod": "Estimated",
      "comment": "estimated via sampling"
    },
    {
      "name": "StorageCosts",
      "costs": {
        "value": 0.146288,
        "unit": "TB",
        "perTimeUnit": "MONTH"
      },
      "computationMethod": "Estimated",
      "comment": "estimated via sampling"
    },
    {
      "name": "MaintenanceCosts",
      "costs": {
        "value": 0,
        "unit": "Credits",
        "perTimeUnit": "MONTH"
      },
      "computationMethod": "Estimated",
      "comment": "Estimated from historic change rate over last ~7 day(s)."
    }
  ]
}
```

Figure 7-2. Example estimate search optimization costs

Note the following when estimating search optimization costs:

- The source table was not enabled for search optimization.

- Investigating search optimization took 24 seconds to process.

- There are three costs:

  - Initial build costs of 10.8 credits.

  - Storage costs of 0.14TB/month, which will vary according to DML operations.

  - Monthly maintenance costs, which will vary according to DML operations.

- Estimates were derived from approximately the past seven days of activity; in this example, zero activity occurred.

As you can see from this information, recent historical DML operations against our target table will affect estimated numbers. Furthermore, the activity period may be affected by the Time Travel setting.

The generated costs are for the table.

I recommend search optimization costs are derived from real-world usage and not from local testing.

After initially enabling SOS, monitor the costs closely.

For every SOS-enabled table and attribute, costs should be monitored on a periodic basis to ensure costs remain within budget appetite.

Enabling Table Search Optimization

Assuming the estimated costs are within the budget appetite and the role is entitled, to enable SOS for an individual table, do this:

```
ALTER TABLE tpc.tpc_owner.lineitem_baseline ADD SEARCH OPTIMIZATION;
```

Enabling search optimization does not imply the Snowflake SOS background service is immediately invoked; you are likely to experience a delay before search optimization is available. To determine whether the Snowflake SOS background service has completed, you must rerun this query:

```
SELECT system$estimate_search_optimization_costs
( 'tpc.tpc_owner.lineitem_baseline' );
```

Figure 7-3 shows the StorageCosts "value" populated indicating the Snowflake SOS background service has completed. A value of 0 indicated the Snowflake SOS background service is in the process of executing.

```
{
    "name": "StorageCosts",
    "costs": {
      "value": 0.145404,
      "unit": "TB"
    },
    "computationMethod": "Measured"
},
```

Figure 7-3. *Search optimization enabled*

You can find more information on search optimization at https://docs. snowflake.com/en/sql-reference/sql/alter-table#label-alter-table-searchoptimizationaction.

You can investigate how search optimization has been applied to a table.

```
DESCRIBE SEARCH OPTIMIZATION ON tpc.tpc_owner.lineitem_baseline;
```

As the partial screenshot shown in Figure 7-4 illustrates, all table attributes are shown as "active" with the method EQUALITY.

expression_id	method	target	target_data_type...	active
1	EQUALITY	L_ORDERKEY	NUMBER(38,0)	true
2	EQUALITY	L_PARTKEY	NUMBER(38,0)	true
3	EQUALITY	L_SUPPKEY	NUMBER(38,0)	true
...				
13	EQUALITY	L_RECEIPTDATE	DATE	true
14	EQUALITY	L_SHIPINSTRUCT	VARCHAR(25)	true
15	EQUALITY	L_SHIPMODE	VARCHAR(10)	true

Figure 7-4. *Table search optimization results*

For reporting purposes you can also extract the "method" and "target" programmatically.

```
SELECT "method",
       "target"
FROM   TABLE ( RESULT_SCAN ( last_query_id()))
WHERE  "active" = 'true';
```

A drawback to implementing table search optimization is cost. Maintaining all attributes on high-velocity, low-volume DML environments results in frequent SOS invocation. We should only consider enabling search optimization for those attributes or partial attributes used within query predicates, discussed next.

Enabling Attribute Search Optimization

Having explained how to enable search optimization for a table, you can now investigate how to set up search optimization for both individual attributes and JSON fields within a VARIANT data type for a table. Most data types are supported, though there are some notable exceptions listed earlier in "Excluded Scenarios."

Three types of attribute search optimization are supported.

- EQUALITY: Match for NUMBER, STRING, BINARY, and VARIANT JSON fields

- SUBSTRING: Partial match for STRING BINARY and VARIANT JSON fields

- GEO: Match for GEOGRAPHY data type

To enable SOS for an individual table attribute, use this:

```
ALTER TABLE tpc.tpc_owner.lineitem_baseline
ADD SEARCH OPTIMIZATION
ON EQUALITY ( l_shipmode );
```

Where attributes meet search optimization criteria, you can set both EQUALITY and SUBSTRING as shown next:

```
ALTER TABLE tpc.tpc_owner.lineitem_baseline
ADD SEARCH OPTIMIZATION
ON EQUALITY ( l_shipmode ), SUBSTRING ( l_shipmode );
```

Now examine active search optimization.

```
DESCRIBE SEARCH OPTIMIZATION ON tpc.tpc_owner.lineitem_baseline;
```

Confirm search optimization is enabled for both methods, as shown in Figure 7-5.

expression_id	method	target	target_data_type	active
1	EQUALITY	L_SHIPMODE	VARCHAR(10)	true
2	SUBSTRING	L_SHIPMODE	VARCHAR(10)	true

Figure 7-5. *Search optimization methods enabled*

You can leave the GEOGRAPHY data type for your further investigation.

You can find more information on search optimization at https://docs.snowflake. com/en/sql-reference/sql/alter-table#label-alter-table-searchoptimizationac tion.

Table Type Support

In this section I will cover the different types of tables supported by SOS. We build each table from the same source table, snowflake_sample_data.tpch_sf1000.lineitem, and then apply the same search optimization criteria before testing the outcome. You then select an arbitrary high-cardinality value for l_partkey used for every query returning 27 rows.

```
USE WAREHOUSE IDENTIFIER ( $tpc_warehouse_xs );

SELECT l_partkey,
       count(*)
FROM   tpc.tpc_owner.lineitem_baseline
GROUP BY l_partkey
HAVING   count (*) > 2
LIMIT 10;
```

The steps are identical for each table type; let's now investigate how SOS works with each table type.

Standard Table

Creating lineitem_baseline_std with a subset of attributes ensures you have a consistent starting point for later query profile comparison. I also summarize DATE attribute l_shipdate to YYYYMM format and convert the data type to VARCHAR.

```
SET tpc_warehouse_XL = 'tpc_wh_xlarge';

USE WAREHOUSE IDENTIFIER ( $tpc_warehouse_xl );
```

Re-creating lineitem_baseline_std will take a few minutes:

```
CREATE OR REPLACE TABLE lineitem_baseline_std
AS
SELECT l_shipmode, l_partkey, l_comment,
       TO_VARCHAR (
         DATE_PART ( YEAR, l_shipdate )||
           DATE_PART ( MONTH, l_shipdate )) AS l_shipdate_yyyymm
FROM snowflake_sample_data.tpch_sf1000.lineitem;
```

To ensure consistency when checking Search Optimization usage, we now create EQUALITY and SUBSTRING search access paths for l_shipmode:

```
ALTER TABLE tpc.tpc_owner.lineitem_baseline_std
ADD SEARCH OPTIMIZATION
ON EQUALITY ( l_shipmode ), SUBSTRING ( l_shipmode ), EQUALITY
( l_partkey );
```

Now examine active search optimization.

```
DESCRIBE SEARCH OPTIMIZATION ON
tpc.tpc_owner.lineitem_baseline_std;
```

Check search optimization is enabled for the expected attributes, as shown in Figure 7-6.

expression_id	method	target	target_data_type	active
1	EQUALITY	L_SHIPMODE	VARCHAR(10)	true
2	EQUALITY	L_PARTKEY	NUMBER(38,0)	true
3	SUBSTRING	L_SHIPMODE	VARCHAR(10)	true

Figure 7-6. *Standard table search optimization enabled*

Estimating search optimization costs for the standard table indicates the SOS is enabled.

```
SELECT system$estimate_search_optimization_costs
( 'tpc.tpc_owner.lineitem_baseline_std' );
```

You should see search optimization is enabled.

Let's now query the table to invoke search optimization by using an enabled attribute within the predicates.

```
SET tpc_warehouse_M = 'tpc_wh_medium';

USE WAREHOUSE IDENTIFIER ( $tpc_warehouse_M );

SELECT l_comment
FROM   tpc.tpc_owner.lineitem_baseline_std
WHERE  l_partkey = '31587234';
```

The query ran in 1.6 seconds.

Checking the query profile as shown in Figure 7-7 proves search optimization was successfully used.

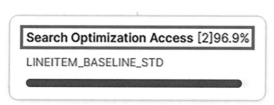

Figure 7-7. *Standard table search optimization profile*

Dynamic Table

Create a dynamic table called dt_lineitem_baseline_sos; note that this will take some time to complete, in my test environment about one hour and 45 minutes. We also summarize DATE attribute l_shipdate to YYYYMM format and convert the data type to VARCHAR:

```
SET tpc_warehouse_XL = 'tpc_wh_xlarge';

USE WAREHOUSE IDENTIFIER ( $tpc_warehouse_xl );

CREATE OR REPLACE DYNAMIC TABLE dt_lineitem_baseline_sos
TARGET_LAG = '30 MINUTES'
WAREHOUSE  = tpc_wh_xsmall
AS
SELECT l_shipmode, l_partkey, l_comment,
       TO_VARCHAR (
         DATE_PART ( YEAR, l_shipdate )||
           DATE_PART ( MONTH, l_shipdate )) AS l_shipdate_yyyymm
FROM    snowflake_sample_data.tpch_sf1000.lineitem;
```

Now resume the dynamic table.

```
ALTER DYNAMIC TABLE dt_lineitem_baseline_sos RESUME;
```

Then refresh the dynamic table.

```
ALTER DYNAMIC TABLE dt_lineitem_baseline_sos REFRESH;
```

Set search optimization on the desired attributes.

```
ALTER TABLE tpc.tpc_owner.dt_lineitem_baseline_sos
ADD SEARCH OPTIMIZATION
ON EQUALITY ( l_shipmode ), SUBSTRING ( l_shipmode ), EQUALITY
( l_partkey );
```

Now examine active search optimization.

```
DESCRIBE SEARCH OPTIMIZATION
ON tpc.tpc_owner.dt_lineitem_baseline_sos;
```

Check that search optimization is enabled for the expected attributes, as shown in Figure 7-8.

expression_id	method	target	target_data_type	active
1	EQUALITY	L_SHIPMODE	VARCHAR(10)	true
2	EQUALITY	L_PARTKEY	NUMBER(38,0)	true
3	SUBSTRING	L_SHIPMODE	VARCHAR(10)	true

Figure 7-8. *Dynamic table search optimization enabled*

However, when we attempt to estimate search optimization costs we will see an error:

```
SELECT system$estimate_search_optimization_costs ( 'tpc.tpc_owner.dt_
lineitem_baseline_sos' );
```

Invalid value ['tpc.tpc_owner.dt_lineitem_baseline_sos'] for function 'SYSTEM$ESTIMATE_SEARCH_OPTIMIZATION_COSTS', parameter 1: argument is not a supported table for search optimization.

While search optimization appears to be set for dynamic tables, you cannot see the costs.

Let's examine the dynamic table dt_lineitem_baseline_sos.

```
SET tpc_warehouse_M = 'tpc_wh_medium';

USE WAREHOUSE IDENTIFIER ( $tpc_warehouse_M );

SELECT l_comment
FROM    tpc.tpc_owner.dt_lineitem_baseline_sos
WHERE   l_partkey = '31587234';
```

The query ran in 8.5 seconds.

Checking the query profile *did not* show search optimization access but instead a table scan, as shown in Figure 7-9.

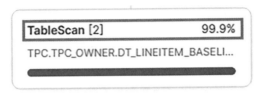

Figure 7-9. *Dynamic table search optimization profile*

You can conclude that search optimization at the time of writing is not fully implemented; the ability to enable SOS indicates it's a work in progress.

Transient Table

I would not usually enable SOS on a transient table though your use case may require SOS enabled for specific transient tables. This section is to prove or disprove that SOS can be enabled for a transient table.

Create a transient table called lineitem_baseline_trans. You also summarize DATE attribute l_shipdate to YYYYMM format and convert the data type to VARCHAR.

```
SET tpc_warehouse_XL = 'tpc_wh_xlarge';

USE WAREHOUSE IDENTIFIER ( $tpc_warehouse_xl );

CREATE OR REPLACE TRANSIENT TABLE lineitem_baseline_trans
AS
SELECT l_shipmode, l_partkey, l_comment,
      TO_VARCHAR (
         DATE_PART ( YEAR, l_shipdate )||
            DATE_PART ( MONTH, l_shipdate )) AS l_shipdate_yyyymm
FROM   snowflake_sample_data.tpch_sf1000.lineitem;
```

Now attempt to add search optimization to the new TRANSIENT table lineitem_
baseline_trans.

```
ALTER TABLE tpc.tpc_owner.lineitem_baseline_trans
ADD SEARCH OPTIMIZATION
ON EQUALITY ( l_shipmode ), SUBSTRING ( l_shipmode ), EQUALITY ( l_
partkey );
```

Now examine active Search Optimization:

```
DESCRIBE SEARCH OPTIMIZATION ON tpc.tpc_owner.lineitem_baseline_trans;
```

Check the search optimization is enabled for the expected attributes, as shown in
Figure 7-10.

expression_id	method	target	target_data_type	active
1	EQUALITY	L_SHIPMODE	VARCHAR(10)	true
2	EQUALITY	L_PARTKEY	NUMBER(38,0)	true
3	SUBSTRING	L_SHIPMODE	VARCHAR(10)	true

Figure 7-10. *Transient table search optimization enabled*

Estimating search optimization costs for the transient table indicates SOS is enabled.

```
SELECT system$estimate_search_optimization_costs ( 'tpc.tpc_owner.lineitem_
baseline_trans' );
```

A JSON record should be returned indicating search optimization is enabled.
Let's examine the transient table lineitem_baseline_trans.

```
SET tpc_warehouse_M = 'tpc_wh_medium';
```

```
USE WAREHOUSE IDENTIFIER ( $tpc_warehouse_M );
```

```
SELECT l_comment
FROM   tpc.tpc_owner.lineitem_baseline_trans
WHERE  l_partkey = '31587234';
```

263

The query ran in 1.7 seconds.

Checking the query profile as shown in Figure 7-11 proves search optimization was successfully used.

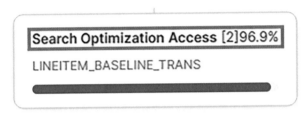

Figure 7-11. *TRANSIENT table search optimization profile*

Temporary Table

I would not usually attempt to enable SOS on a temporary table. This section is to prove or disprove SOS can be enabled for a temporary table. Note the keywords TEMP and VOLATILE are synonyms for TEMPORARY.

Create a temporary table called lineitem_baseline_temp. You can also summarize DATE attribute l_shipdate to YYYYMM format and convert data the type to VARCHAR:

```
SET tpc_warehouse_XL = 'tpc_wh_xlarge';

USE WAREHOUSE IDENTIFIER ( $tpc_warehouse_xl );

CREATE OR REPLACE TEMPORARY TABLE lineitem_baseline_tmp
AS
SELECT l_shipmode, l_partkey, l_comment,
      TO_VARCHAR (
         DATE_PART ( YEAR, l_shipdate )||
            DATE_PART ( MONTH, l_shipdate )) AS l_shipdate_yyyymm
FROM   snowflake_sample_data.tpch_sf1000.lineitem;
```

Now attempt to add search optimization to the new TEMPORARY table lineitem_baseline_tmp.

```
ALTER TABLE tpc.tpc_owner.lineitem_baseline_tmp
ADD SEARCH OPTIMIZATION
ON EQUALITY ( l_shipmode ), SUBSTRING ( l_shipmode ), EQUALITY
( l_partkey );
```

You cannot add search optimization to temporary tables and should see the following error:

> **_Error: Invalid materialized view definition. Source table 'TPC.TPC_OWNER.LINEITEM_BASELINE_TMP' should not be temporary. (line nn)_**

You cannot enable SOS for temporary tables.

Conclusion

Search optimization cannot be enabled for all table types, as shown in Figure 7-12.

	Standard	Dynamic	Transient	Temporary
Can be SOS Enabled?	Yes	Yes	Yes	No
Query uses SOS?	Yes	No	Yes	No

Figure 7-12. *Search optimization table coverage*

Rerunning the standard table query after our dynamic table had been created did not cause the dynamic table to be used. Search optimization was preferred.

```
SELECT l_comment
FROM   tpc.tpc_owner.lineitem_baseline_std
WHERE  l_partkey = '31587234';
```

Disabling Table Search Optimization

You can disable SOS by doing the following:

```
ALTER TABLE tpc.tpc_owner.lineitem_baseline_std DROP SEARCH OPTIMIZATION;
```

Note SOS will report StorageCosts after SOS is disabled; this is consistent with Time Travel and Fail Safe behavior where you would expect search optimization to be preserved in a consistent manner with object content.

```
SELECT system$estimate_search_optimization_costs ( 'tpc.tpc_owner.lineitem_
baseline_std' );
```

Confirm search optimization is disabled, as shown in Figure 7-13.

```
"tableName": "LINEITEM_BASELINE_STD",
"searchOptimizationEnabled": false,
```

Figure 7-13. *Search optimization disabled*

Timeliness

As previously stated, the Snowflake SOS background service runs asynchronously; therefore, I cannot be sure of when the search access paths for an enabled table will complete. Furthermore, when the table data changes, the search access paths will update during which time queries might run more slowly. This behavior is explained within the Snowflake documentation at https://docs.snowflake.com/en/user-guide/search-optimization-service#how-the-search-optimization-service-works.

I suggest search optimization should not be implemented on empty tables nor on tables where high velocity DML operations occur.

Best Practices

Implementing SOS can significantly improve the performance of some queries. I offer the following guidelines when considering SOS:

- The candidate table should have a high number of micro-partitions.

- High cardinality attributes are ideally suited to SOS.

- Apply SOS on a selective attribute basis where known access paths exist.

- Use SOS for tables with low-velocity DML operations.

- Match existing point lookup query predicates to attribute values.

Conversely, these scenarios are to be avoided when implementing SOS:

- Low number of micro-partitions

- Low cardinality attributes

- Unsupported data types

- Low query execution time (less than a few seconds)

- The query predicates are:

  - Not `EQUALITY` or `SUBSTRING`

  - Contained within an `IN` list

  - The result of a subquery

- A relatively large result set in comparison to the full data set

- A SQL function in the query on the target table attribute

Search optimization does not support the leading attribute of a cluster key as this already provides for micro-partition pruning.

Snowflake is continually evolving, and SOS is no exception. At the time of writing, query join acceleration, promising a dramatic speedup for star schema joins, is being worked on, as disclosed during Snowflake Summit in June 2023.

Summary

This chapter introduced SOS and then suggested where SOS may best be deployed within a typical application footprint before indicating optimal usage and limitations.

SOS is not a silver bullet and must be selectively applied. SOS implemented on tables with high-velocity, low-volume DML operations will prove costly to maintain. I also showed that implementing SOS at the table level in most scenarios will also prove costly, and you should prefer to implement SOS for individual attributes instead. However, also note that SOS provides a capability to reference individual values within a JSON record, which is a very useful feature.

Your focus should be on enabling SOS for individual attributes only.

The walk-through of search optimization provided insight into how SOS works, noting the asynchronous nature of the service. Both compute costs and storage costs are incurred, noting the interaction with the Time Travel and Fail Safe settings. I explained how to investigate both storage and compute costs then working through reference test cases for standard tables, dynamic tables, transient tables, and temporary tables.

Testing is crucial to both proving performance benefits and for controlling costs.

No investigation would be complete without discussing the timeliness of search access path maintenance and the implications of high-velocity DML changes to the source table.

Finally, we offer a best-practice guide summarizing the optimal patterns where SOS provides the most benefit.

For your further investigation, you may find these articles helpful:

- `https://community.snowflake.com/s/article/Search-Optimization-When-How-To-Use`

- `https://community.snowflake.com/s/article/Search-Optimization-When-How-To-Use-Part-2`

- `https://community.snowflake.com/s/article/Search-Optimization-When-How-To-Use-Part-3`

Drawing our investigation into SOS to a close, you will now investigate how to improve the data pipeline processing speed by parallelizing your code.

CHAPTER 8

Parallelization

This chapter marks a change of focus for the remainder of this book as we will now look to solving real-world performance issues. My experience is derived from developing and implementing pragmatic solutions to seemingly intractable problems.

Throughout this chapter, you will investigate a performance issue step-by-step using an example data ingestion process based upon my real-world experience. I will offer a diagnostic approach to educate and inform how to approach problems with the expectation that the implementation of proactive monitoring will expose future risks.

Snowflake is the latest and greatest data warehouse to hit the market and has rightly attracted a lot of positive attention. However, for some, Snowflake is expensive to run; this criticism is mostly ill-founded and arises due to misunderstanding and misapplying best practices when implementing Snowflake. There is also a general reluctance to tune existing applications when porting to Snowflake, a bad mistake to make.

Operating in a cloud-based global marketplace presents different challenges for both data distribution to closed local applications operating within a corporate network and service delivery via dedicated on-prem hardware in fixed data centers. Your approach must adapt, because what has served you well in the past will not serve you well into the future.

Curation of data incurs both cost and time. In our new global marketplace paradigm, you must also replicate data seamlessly and accept that replicating data incurs both cost and time.

While you adopt a Snowflake-centric view of this data marketplace strategy, you must also consider how to act as a "data master" and enable distribution to other vendor offerings.

There is plenty to investigate, so let's look at some foundational information to expose where problems occur in some existing applications. Then later you'll see how to remediate them.

© Andrew Carruthers 2024
A. Carruthers, *Tuning the Snowflake Data Cloud*, https://doi.org/10.1007/979-8-8688-0379-6_8

Foundational Information

In this section, I will describe the basic concepts relating to a typical application design used within previous chapters while expanding the application scope to distribute data globally.

For this investigation let's will assume an additional requirement to distribute our offerings globally, as indicated by Figure 8-1.

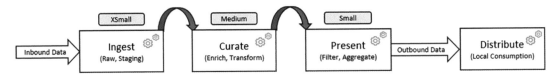

Figure 8-1. *Example global application perspective*

The example focuses on batch inbound data, while the same principles apply to inbound streamed data. The velocity is likely to be higher, and the volume of each streamed data set will be lower. The net effect of ingesting streamed data will be higher micro-partition churn. To mitigate against performance impact, you might adopt a halfway house of consolidating streamed data into batches before applying in bulk.

What follows is a broad outline of every application at a very high level. Data is ingested on the left, and products are consumed on the right. Extending our consumption model by distributing data across marketplaces, regions, cloud service providers (CSPs), and disparate platforms presents further challenges.

IYou offer this information as a broad analog for all application data flows along with indicative Data Manipulation Language (DML) actions. The information supplied is not intended to suggest there is a single right way to construct applications.

Let's investigate the outline function of each application container individually.

Data Products

Throughout this chapter I use the term "*data product*" to refer to data curated into a product or feature resulting in a value-added component that a client consumes. Data products may be distributed free of charge or commercialized in some manner. In essence, a data product is "something" a client wants to consume.

Ingest

Apart from the initial seeding of an application, steady-state data ingestion typically involves modest data volumes at known frequency. You expect your load testing will have informed your warehouse strategy and provide an indicative maximum velocity per feed or consistent data set ingestion approach.

Ingesting data into application raw or staging tables is a prerequisite before merging ingested data into a suite of core tables where you curate your data products. As a rule of thumb and discussed at length in Chapter 6, you should plan for X-Small warehouses for data ingestion.

Raw or staged data contains the following:

- New data that does not exist in the core tables; new data is usually inserted into core tables.

- Old data marked for removal from the core tables; old data is usually physically deleted or logically deleted from core tables.

- Changed data for existing records in the core tables; changed data is usually updated, or new data is inserted, and then old data is logically deleted from core tables.

You should know both the frequency of data ingestion (the velocity) and the type of DML (the volume) for `INSERT`, `UPDATE`, and `DELETE` operations to be performed on our core tables as this information will prove essential later.

All applications ingest data. For the purposes of our investigation, you will use the supplied TPC data set as our data source.

Curate

Data products are created from the combination of intellectual property usually in the form of bespoke logic and the ingestion of data. When data meets business process, value results.

As I identified in my earlier book *Building the Snowflake Data Cloud*, data in its correct context provides information. This is more clearly stated using the data, information, knowledge, wisdom (DIKW) pyramid; see `https://en.wikipedia.org/wiki/DIKW_pyramid`.

Information is derived from data via consolidation, cleansing, transformation, and consumption processes. Knowledge—the intellectual property—is derived from information, and wisdom is gained from applying knowledge. Figure 8-2 illustrates the relationship between each layer of the DIKW pyramid, demonstrating the value chain.

***Figure 8-2.** DIKW pyramid*

Data products are the resultant "value-add" enabling organizations to monetize their intellectual property without disclosing their internal methodology, or, if you prefer, their "secret sauce."

With intellectual property embedded within the data processing pipelines, curating data products involves maintaining core table data with content from the raw or staging tables. As a rule of thumb and discussed at length in Chapter 6, you should plan for a Medium warehouse for data curation. However, as you will see later within this chapter, parallelization may allow a smaller warehouse to be used in a more efficient manner.

Without igniting a debate regarding the data modeling style implemented within an application, you should understand the profile of your core data and the cluster keys and match your ingestion process to the core data model and structures. The efficient processing of data is the core theme of this chapter, and I will unpack this theme in detail later.

Produce

Your objective is to deliver curated data products to your clients. Traditional data distribution mechanisms such as secure file transfer via SFTP and on-premise dedicated server provisioning are being replaced with automated, seamless approaches. Along with supporting traditional data distribution mechanisms, Snowflake offers several new and innovative data distribution approaches, as shown in Figure 8-3.

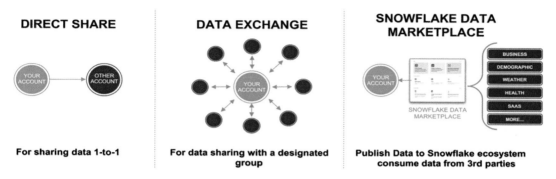

Figure 8-3. *Snowflake data distribution patterns*

In addition to those patterns shown in Figure 8-3, you might also deliver your data products via external mechanisms, but for the purposes of this section you will limit yourself to local Snowflake account reporting or Secure Direct Data Sharing (SDDS) where you implement point-to-point data sharing, as shown in Figure 8-4.

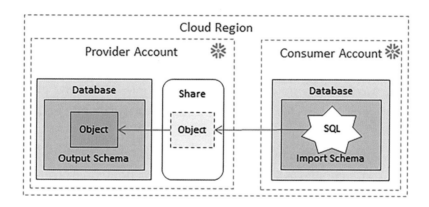

Figure 8-4. *Secure Direct Data Share usage*

Figure 8-4 shows the capability to share the current version of data held within an object, or a functional component with a co-located client account. The important point to note for SDDS is that both the provider and consumer accounts must reside within the same CSP and region. Replication can be to any supported CSP and region, noting that replication incurs additional cost.

Data products are not delivered to clients as an "all-or-nothing" proposition. In other words, not every client will purchase all licensable data products; they may prefer to purchase a subset of available data products instead. You must consider how to entitle your data products for client consumption. Due to both complexity and performance considerations, this subject is treated separately within the next chapter.

You may choose a variety of data distribution models and do not prescribe any particular approach. For local Snowflake account consumption, as a rule of thumb and discussed at length in Chapter 6, you should plan for Small warehouses for data consumption.

Distribution Venues

Your focus within this book so far has been to develop an understanding of how Snowflake works internally to minimize costs and maximize performance. It might not be obvious why you are investigating data product distribution across disparate venues, CSP locations, and software platforms.

Using Snowflake to master data product curation is an excellent strategy for success though mastering typically occurs in a single location, preferably close to the inbound data sources. But your clients probably operate globally and want to consume your data products according to their individual needs, which may involve specific file formats, data subsets, geographical locations, and alternate consumption products and platforms.

Any data product provider must consider their offering as one or more inputs to their client's infrastructure, perhaps a single box within a complex environment. Adopting a client-centric approach provides insight and informs your approach to data product distribution.

Data replication can be both more complex and costly than first thought. Efficient processing of the inbound data will also positively impact how you distribute the data products too.

To provide broader context for the later investigation, let's briefly examine some distribution venues.

Snowflake Marketplaces

Snowflake operates both Private Listings and Marketplace, and each is maintained through Snowsight, the default Snowflake-supplied user interface. Snowsight enables client access to predefined objects within a Snowflake account.

Figure 8-5 illustrates the Snowflake-provisioned one-to-many models for data interchange between a single provider and one or more consumers.

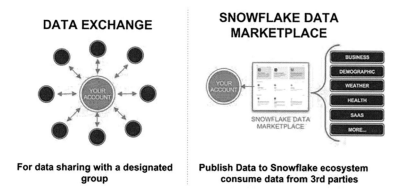

Figure 8-5. *Data exchange and Marketplace*

Data exchange and Marketplace provision data products within a single CSP region and across CSPs and regions providing global coverage to all Snowflake locations.

Snowflake Regions and CSPs

Snowflake data product distribution occurs via replication, that is, a timed event to ingest changed micro-partitions from a master site into a secondary site. Using replication incurs latency. There may be time gaps between mastering and availability of a data product at a source site and fulfilment at a replicated site.

Figure 8-6 illustrates Snowflake share and replication options.

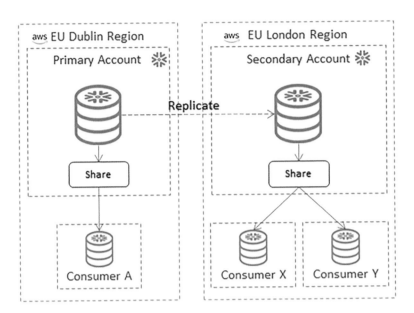

Figure 8-6. *Snowflake data interchange*

You can find more information about Snowflake-supported regions at `https://docs.snowflake.com/en/user-guide/intro-regions`.

When implementing data sharing across regions and CSPs, you must be mindful of costs. You can find more information on Snowflake data transfer costs at `https://docs.snowflake.com/en/user-guide/cost-understanding-data-transfer?utm_source=legacy`.

Snowflake replication costs are always paid by the data provider regardless of the data transfer mechanism.

Snowflake does not permit re-sharing either shared or replicated data, as Figure 8-7 illustrates.

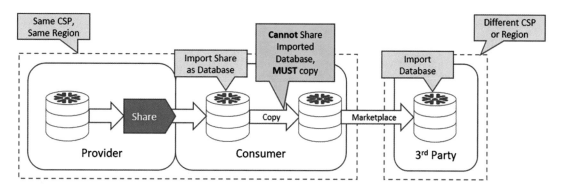

Figure 8-7. *Snowflake data share restriction*

Imported databases, whether derived from a share or from replicated database, cannot be reshared. The same principle applies to data products shared via data exchange and marketplace.

CSP Marketplaces

CSPs also operate their own marketplaces, for example, AWS Data Exchange (ADX), Azure Marketplace, and Google Cloud Marketplace. Data interchange costs must also be considered for hydrating each target marketplace. Note that data egress costs may apply.

Iceberg, Platforms, and S3-Compatible Storage Support

Snowflake supports Iceberg tables; see `https://www.snowflake.com/blog/unifying-iceberg-tables/`.

Outside of the Snowflake ecosystem, several other distribution venues exist including Google Big Query, Databricks, Microsoft Fabric, and other non-Snowflake supported CSPs.

You should also be mindful that Snowflake offers third-party data integration capability via S3-compatible storage.

Logging

Multiple processes that are logging event information into a single table will serialize all concurrent processes as each logging process locks the target table and micro-partitions are written.

I offered a solution to serialized logging in Chapter 6 by implementing an EVENT table and noted that only a single EVENT table can be active at any given time.

EVENT information must be periodically collated into a separate log table for long-term audit trail preservation.

Optimizing Data Processing

Every application has an optimal or target processing time from data landing in raw or staging tables through to content appearing within the client-consumed data product. In most cases, clients will pay a premium for faster data product updates, change propagation, and availability.

You goal should be to both reduce cost and improve timeliness across the whole application life cycle from the initial point of data ingestion through to client consumption. For this to become reality, you must adopt a holistic approach to identifying the root causes of both cost consumption and latency and then apply effective remediation.

I assume you have checked your code for Cartesian joins, long compilation times, long execution times, and long table scans as part of the user acceptance tests and commissioning into production.

Let's look at an example system as viewed from the client perspective.

Problem Statement

Several clients have observed that the time it takes for data to appear in their licensed product is getting progressively slower month over month. The first step is to validate the client's claim by checking the telemetry information logged for the feed. Let's assume the feed contains normal data volumes, and the logged information corroborates the client claim. You confirm the feed runtimes have marginally degraded over time.

From the information supplied and analysis conducted by your product support team, you deduce one single inbound data feed, and onward ingestion into the core data set is affected. Figure 8-8 illustrates the left-to-right data flows for our sample feed.

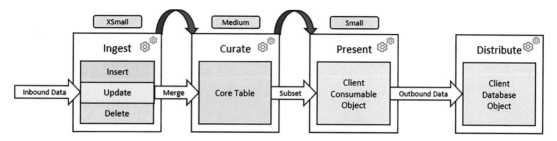

Figure 8-8. *Sample data feed*

Warehouse Factors

The investigation starts by identifying factors external to the feed process that may affect data ingestion, starting with the warehouse. In this example and typical for most data ingestion patterns, you will use an X-Small warehouse.

Answer these questions:

- Are there any warehouse queueing, spills to disk, or OOMs evident?

- Is the warehouse overloaded, queueing, or blocking?

- Is the data feed overrunning its schedule leading to feeds backing up?

- Are costs increasing over time?

Figure 8-9 illustrates the warehouse scaling options.

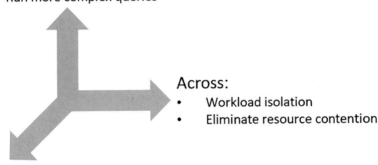

Figure 8-9. *Warehouse scaling options*

For the affected data load, you should first look at the latest run query profile. You are looking for spills to disk and OOM errors, both covered in Chapter 3. If either is evident, you should consider either scaling up the ingest warehouse size to the next size or reducing warehouse concurrency.

The following query repeated from Chapter 3 identifies spills to disk and OOM errors. Replace warehouse_name with your chosen value:

```
SET tpc_owner_role    = 'tpc_owner_role';
SET tpc_warehouse_XS  = 'tpc_wh_xsmall';
SET tpc_warehouse_XL  = 'tpc_wh_xlarge';
SET tpc_database      = 'tpc';
SET tpc_owner_schema  = 'tpc.tpc_owner';

USE ROLE      IDENTIFIER ( $tpc_owner_role   );
USE DATABASE  IDENTIFIER ( $tpc_database     );
USE SCHEMA    IDENTIFIER ( $tpc_owner_schema );
USE WAREHOUSE IDENTIFIER ( $tpc_warehouse_xs );
```

```
SELECT query_id,
       warehouse_name,
       warehouse_size,
       bytes_spilled_to_local_storage,
       bytes_spilled_to_remote_storage,
       bytes_sent_over_the_network
FROM   snowflake.account_usage.query_history
WHERE  warehouse_name                       = '<YOUR_WAREHOUSE_HERE>'
AND    bytes_spilled_to_remote_storage > 0;
```

The next check is for concurrency; you should investigate the number of concurrent processes running at the same time your clients were reporting issues. If you observe either queueing or blocking, then you should consider scaling out your ingest warehouse by adding clusters. You could also implement the Query Acceleration Service as discussed in Chapter 6.

The following query repeated from Chapter 6 identifies overlapping subset of records using the same named warehouse and specific date_time. Replace variables with your chosen values:

```
CREATE OR REPLACE VIEW v_warehouse_workload_by_hour COPY GRANTS
AS
SELECT warehouse_name,
       start_time,
       end_time,
       query_id,
       query_text,
       total_elapsed_time / 1000  AS total_elapsed_time_in_secs,
       queued_overload_time ,
       transaction_blocked_time,
       DATE_PART (          'YYYY', start_time )||
       LPAD ( DATE_PART ( 'MM',   start_time ), 2, '0' )||
       LPAD ( DATE_PART ( 'DD',   start_time ), 2, '0' )||'_'||
       LPAD ( DATE_PART ( 'HOUR', start_time ), 2, '0' )
                                  AS date_time
FROM   snowflake.account_usage.query_history
WHERE  execution_time <> 0
```

```
ORDER BY warehouse_name,
         start_time DESC;

SELECT v1.start_time,
       v1.end_time,
       v1.query_id,
       v1.total_elapsed_time_in_secs,
       v1.date_time
FROM    v_warehouse_workload_by_hour v1
WHERE   EXISTS
        (
        SELECT 1
        FROM    v_warehouse_workload_by_hour v2
        WHERE   v2.start_time <= v1.end_time
        AND     v2.end_time   >= v1.start_time
        AND     v2.date_time   = v1.date_time
        AND     v2.query_id   != v1.query_id
        )
AND     v1.warehouse_name = '<YOUR_WAREHOUSE_HERE>'
AND     v1.date_time      = '<YOUR_DATE_TIME_HERE>'
ORDER BY v1.start_time DESC;
```

Scaling across should be avoided as this approach can both reduce concurrency and increase costs, as explained within Chapter 6.

Next, check the logged information to ensure feed runtimes do not exceed the service-level agreements (SLAs), and ensure each run completes before the next batch cycle. You must prove there is no backlog accruing throughout the day, which unwinds during the quiet times. Historical performance monitoring will prove very useful in predicting future capacity issues, because trends often foretell future problems. Assuming event logging is used, extracting the start time from the end time for each process will determine the runtime. This is an exercise I leave to you for your further investigation. Note that Chapter 6 contains information about event logging.

Having investigated the external factors that may have affected your feed, you determine the warehouse is not overloaded and shows no signs of queueing. Analyzing the recent query profiles does not show spills to disk or OOMs. The logged information does not show feed runtimes or breach the SLAs, but historical performance monitoring does show a worrying trend: feed runtimes are increasing for a steady workload profile.

You should also check costs where you find warehouse runtimes are increasing and therefore increasing consumption costs. Then you notice something surprising: data replication costs are higher than the cost of curating data.

Ingest Factors

Let's continue the investigation by identifying factors internal to the feed process that may affect data ingestion.

The inbound data lands in the raw or staging table. The actual delivery mechanism is unimportant, but to add flavor, let's assume the bulk load operation occurs via a COPY command from file held in external CSP storage.

Raw or staged data contains new records for INSERT, changed records for UPDATE, and old data marked for DELETE. You assume either unique records are loaded into our staging table or a mechanism exists to de-duplicate data prior to MERGE into the target core table.

The ingest process must also identify a primary key, unique composite key, or hash for merging into the target core table. Your real-world application will already have solved these challenges, but new applications will need to consider how to do the same.

Not shown are validation routines and setting NULL values to known defaults, these may be edge cases and either positively excluded from your application design or explicitly included within your application design.

Figure 8-10 shows the steps involved in preparing the feed for ingestion into a core table.

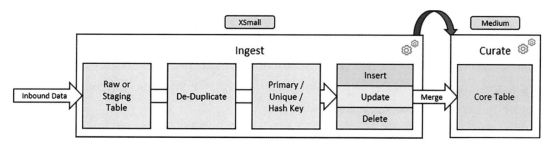

Figure 8-10. *Ingest data preparation*

Having set out contextual information, let's aim to answer these questions:

- DML volume for INSERT, UPDATE, and DELETE operations

- Core table data volume and number of micro-partitions

- Core table cluster key definition; you can assume a cluster key
  is defined

- Staged table data profile

- Core table data profile

You assume the raw or staged data contains equal numbers of INSERT, UPDATE, and DELETE operations. From Chapter 4 you know that both INSERT and DELETE operations complete faster than UPDATE operations.

Knowing the target core table data volume allows you to determine the percentage change for INSERT, UPDATE, and DELETE operations. The number of target core table micro-partitions is useful but not essential; of more significance is the number of micro-partitions to be replicated.

The staged data profile is of particular interest, you can assume the sample data is typical in profile; in other words, the data is not skewed nor misrepresenting the usual data loaded.

For the example, you can assume the source staged table data has these attributes of significance for merging data:

- **Unique identifier:** The primary, unique, or hash key for the record

- **Record start date:** The business date from which the record is valid

- **DML operation:** Letter (I, U, D) indicating the type of operation to
  perform on the target core table

The three attributes identified from the source staged table data enable us to effectively implement a MERGE statement. But the three attributes are highly unlikely to match the target table cluster key, which must be defined according to business needs, not technical data maintenance needs.

You must understand how the staged data matches the target core table data. Note that the cluster key will lead with the least selective attribute first and then the next least selective attribute, and so on. This information is crucial for developing the parallelization strategy.

Curation Factors

You might ask yourself why you cannot separate the three INSERT, UPDATE, and DELETE parts of the MERGE statement and run these in parallel. Answering this question involves understanding how Snowflake maintains micro-partitions and implements table locking; I discussed these subjects within Chapter 4.

Figure 8-11 shows the impact of attempting to run three concurrent processes for INSERT, UPDATE, and DELETE in parallel against a single-core table. You will experience blocking as the first process locks the core table and completes a DELETE operation; then the next queued process will lock the core table and complete before the final process locks the core table and completes.

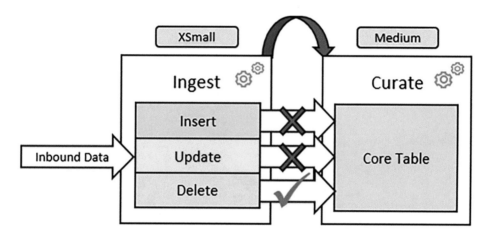

Figure 8-11. *Serialized DML operations*

The "Law of Unintended Consequences" (https://en.wikipedia.org/wiki/Unintended_consequences) serializes the parallel processes. The observant might note this is exactly the same effect described for logging information in Chapter 6.

If you are determined to serialize the DML operations, the optimal order of application is as follows:

- DELETE: Reduces the data volume for later DML activity

- UPDATE: Operates against the minimal data volume in the object

- INSERT: Adds new records to increase object data volume

Merging data may also include enriching with reference data and maintenance of bitemporal attributes mentioned for completeness only.

If your data model is "Insert only," then you might avoid some concurrency issues, but most applications are not designed for "Insert only" at the outset.

Your aim is to maximize parallelization while minimizing cost; I discuss this next.

Parallel Processing

Every time you execute DML, you either instantiate a new warehouse or consume a processing unit from an already running warehouse. Figure 8-12 illustrates the effect of instantiating a new X-Small warehouse for a single DML statement.

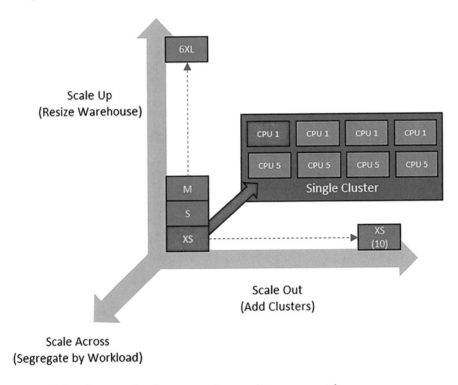

Figure 8-12. *Warehouse single processing unit consumption*

Your aim is to utilize every processing unit within every active cluster. When a warehouse runs, you pay for the whole runtime; costs are not apportioned to each processing unit executing DML. For maximum efficiency, you must utilize every processing unit within active warehouses, as shown in Figure 8-13.

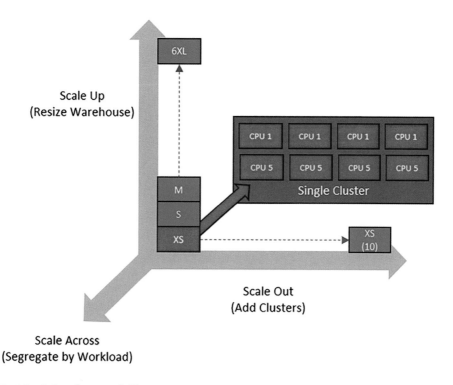

Figure 8-13. *Warehouse full processing unit consumption*

As I identified previously, splitting the MERGE statement into its component DML operations will not work as the outcome serializes the workflow. You need a different approach: segment the target core table.

To parallel process a data load, you must consider these factors:

- How to shard the target core table into physical partitioned tables

- Number of warehouse concurrent processing units required per feed

- Orchestrating physical partition loads

- Impact of partition load completion versus full table load

- Denormalizing physical partitions to represent the physical core table

Figure 8-14 shows a high-level design to implement parallel processing of a single feed noting there will be "n" segmented tables according to the segmentation key chosen.

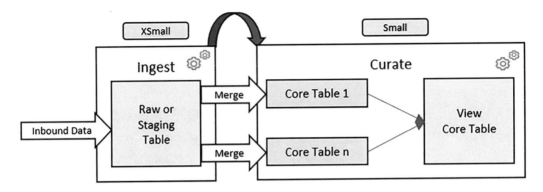

Figure 8-14. *Parallel processing high-level design*

Adopting this design pattern for an existing application allows the selective replacement of a poorly performing ingestion pipeline with highly performant components and minimal system impact.

Setting Up Application Tables

In this section I will simulate an existing application raw or staging table along with a target core table both populated with sample data. Figure 8-15 illustrates the immediate objective.

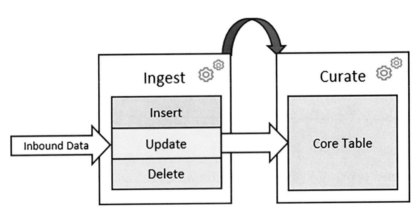

Figure 8-15. *Setup objective*

Begin by declaring a single-core table clustered in a geographic region and use an X-Large warehouse due to high data volumes.

```
USE WAREHOUSE IDENTIFIER ( $tpc_warehouse_xl );
```

Use a standard table called base_customer_order to hold a baseline set of data for later use.

```
CREATE OR REPLACE TABLE base_customer_order
AS
SELECT c.c_custkey    AS customer_key,
       c.c_name       AS customer_name,
       o.o_orderkey   AS order_key,
       o.o_orderdate AS order_date,
       n.n_name       AS nation_name,
       r.r_name       AS region_name
FROM   snowflake_sample_data.tpch_sf1000.region   r,
       snowflake_sample_data.tpch_sf1000.nation   n,
       snowflake_sample_data.tpch_sf1000.customer c,
       snowflake_sample_data.tpch_sf1000.orders   o
WHERE  c.c_custkey   = o.o_custkey
AND    c.c_nationkey = n.n_nationkey
AND    n.n_regionkey = r.r_regionkey;
```

Create a target core table to simulate an existing application table.

```
CREATE OR REPLACE TABLE core_customer_order
AS
SELECT customer_key, customer_name, order_key, order_date,
       nation_name, region_name
FROM   base_customer_order;
```

Now add a clustering key on region_name and nation_name.

```
ALTER TABLE tpc.tpc_owner.core_customer_order
CLUSTER BY ( region_name, nation_name );
```

Reset the warehouse to X-Small.

```
USE WAREHOUSE IDENTIFIER ( $tpc_warehouse_xs );
```

Count the number of records created and identify the latest order_date and highest order_key values, because you will use this information shortly when creating raw or staging data.

```
SELECT count(1),
       MAX ( order_date ),
       MAX ( order_key  )
FROM   core_customer_order;
```

The new table should contain 1,500,000,000 records with the latest order_date of 1998-08-02 and highest order_key value set to 6,000,000,000.

Let's now create a raw or staging table noting the addition of the operation and stg_timestamp attributes.

```
CREATE OR REPLACE TABLE stg_customer_order
(
customer_key    NUMBER(38,0),
customer_name   VARCHAR(25),
order_key       NUMBER(38,0),
order_date      DATE,
nation_name     VARCHAR(25),
region_name     VARCHAR(25),
operation       VARCHAR(1),
stg_timestamp   TIMESTAMP_NTZ
);
```

LIMIT returns random rows. For repeatable test cases with known values, you must create repeatable consistent records for the INSERT operation.

```
CREATE OR REPLACE TABLE base_customer_order_insert
(
customer_key    NUMBER(38,0),
customer_name   VARCHAR(25),
order_key       NUMBER(38,0),
order_date      DATE,
nation_name     VARCHAR(25),
region_name     VARCHAR(25),
operation       VARCHAR(1),
stg_timestamp   TIMESTAMP_NTZ
)
AS
```

```
SELECT customer_key,
       customer_name,
       order_key + 1000000000,
       DATE_TRUNC ( 'DAY', current_date()),
       nation_name,
       region_name,
       'I',
       current_timestamp()
FROM   base_customer_order
WHERE  order_key > 5000000000
LIMIT  100000;
```

Create records for the UPDATE operation.

```
CREATE OR REPLACE TABLE base_customer_order_update
(
customer_key    NUMBER(38,0),
customer_name   VARCHAR(25),
order_key       NUMBER(38,0),
order_date      DATE,
nation_name     VARCHAR(25),
region_name     VARCHAR(25),
operation       VARCHAR(1),
stg_timestamp   TIMESTAMP_NTZ
)
AS
SELECT customer_key,
       customer_name,
       order_key,
       DATE_TRUNC ( 'DAY', current_date()),
       nation_name,
       region_name,
       'U',
       current_timestamp()
FROM   base_customer_order
WHERE  order_key < 1000000000
LIMIT  100000;
```

Create records for a DELETE operation. Note that the important attributes are order_key and operation. The remainder will not be used for this example. However, if logically deleting from a bitemporal model, then stg_timestamp would be used to set the record valid_to date.

```
CREATE OR REPLACE TABLE base_customer_order_delete
(
customer_key    NUMBER(38,0),
customer_name   VARCHAR(25),
order_key       NUMBER(38,0),
order_date      DATE,
nation_name     VARCHAR(25),
region_name     VARCHAR(25),
operation       VARCHAR(1),
stg_timestamp   TIMESTAMP_NTZ
)
AS
SELECT customer_key,
       customer_name,
       order_key,
       DATE_TRUNC ( 'DAY', current_date()),
       nation_name,
       region_name,
       'D',
       current_timestamp()
FROM    base_customer_order
WHERE   order_key BETWEEN 1000000000 AND 5000000000
LIMIT   100000;
```

Using our base data, now populate the raw or staging table, stg_customer_order.

```
INSERT OVERWRITE INTO stg_customer_order
SELECT *
FROM    base_customer_order_insert
UNION ALL
SELECT *
FROM    base_customer_order_update
```

```
UNION ALL
SELECT *
FROM   base_customer_order_delete;
```

Note UNION ALL runs faster than UNION thus avoiding a SORT operation to determine distinct rows.

Let's confirm the raw or staging data has been created as expected:

```
SELECT count(1),
       operation
FROM   stg_customer_order
GROUP BY operation;
```

You should see 100,000 records each for INSERT, UPDATE, and DELETE.

With a low record sample size, you run the risk of generating a skewed data set. To prevent this scenario, confirm you have records for all regions, and note the number of regions for later use.

```
SELECT count(1),
       region_name
FROM   stg_customer_order
GROUP BY region_name;
```

Figure 8-16 shows sample expected results. Because of the way LIMIT works, your results will vary.

COUNT(1)	REGION_NAME
12024	ASIA
15654	EUROPE
10977	AFRICA
66145	AMERICA
195200	MIDDLE EAST

Figure 8-16. *Regions and row counts*

With the raw or staged data created and core table prepared, you are ready to investigate how to modify the application schema.

Testing Core Table Load

With your newly create raw or staged data, let's establish how long your data pipeline takes to merge the content into the target core table.

Commence testing using an X-Small warehouse and increase the size to provide indicative timings having reset the target core table and set the same cluster key for each run.

```
SET tpc_warehouse_XS = 'tpc_wh_xsmall';
SET tpc_warehouse_S  = 'tpc_wh_small';
SET tpc_warehouse_M  = 'tpc_wh_medium';
SET tpc_warehouse_L  = 'tpc_wh_large';
SET tpc_warehouse_XL = 'tpc_wh_xlarge';

USE WAREHOUSE IDENTIFIER ( $tpc_warehouse_XS );
USE WAREHOUSE IDENTIFIER ( $tpc_warehouse_S  );
USE WAREHOUSE IDENTIFIER ( $tpc_warehouse_M  );
USE WAREHOUSE IDENTIFIER ( $tpc_warehouse_L  );
USE WAREHOUSE IDENTIFIER ( $tpc_warehouse_XL );

MERGE INTO core_customer_order c
USING stg_customer_order       s
ON    c.order_key = s.order_key
WHEN     MATCHED AND operation = 'D' THEN DELETE
WHEN     MATCHED AND operation = 'U' THEN
   UPDATE SET customer_key  = s.customer_key,
              customer_name = s.customer_name,
              order_date    = s.order_date,
              nation_name   = s.nation_name,
              region_name   = s.region_name
WHEN NOT MATCHED AND operation = 'I' THEN
   INSERT ( customer_key,
            customer_name,
            order_key,
```

```
            order_date,
            nation_name,
            region_name )
   VALUES ( s.customer_key,
            s.customer_name,
            s.order_key,
            s.order_date,
            s.nation_name,
            s.region_name );
```

Figure 8-17 shows the effect of changing the warehouse size for the same workload MERGE into a consistent reset target core table.

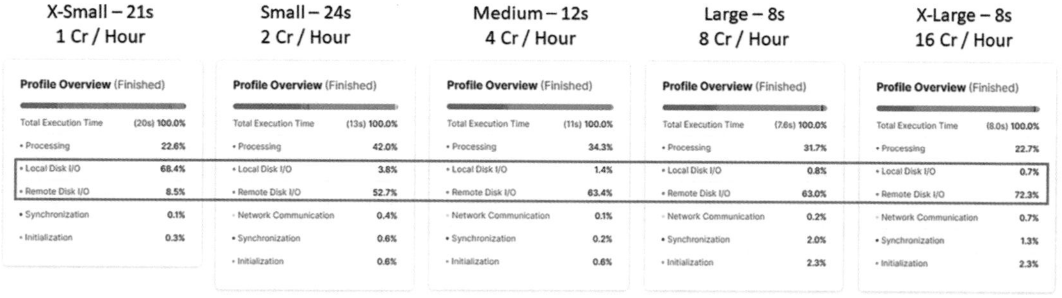

X-Small – 21s 1 Cr / Hour	Small – 24s 2 Cr / Hour	Medium – 12s 4 Cr / Hour	Large – 8s 8 Cr / Hour	X-Large – 8s 16 Cr / Hour
Profile Overview (Finished)	**Profile Overview** (Finished)	**Profile Overview** (Finished)	**Profile Overview** (Finished)	**Profile Overview** (Finished)
Total Execution Time (20s) 100.0%	Total Execution Time (13s) 100.0%	Total Execution Time (11s) 100.0%	Total Execution Time (7.6s) 100.0%	Total Execution Time (8.0s) 100.0%
• Processing 22.6%	• Processing 42.0%	• Processing 34.3%	• Processing 31.7%	• Processing 22.7%
• Local Disk I/O 68.4%	• Local Disk I/O 3.8%	• Local Disk I/O 1.4%	• Local Disk I/O 0.8%	• Local Disk I/O 0.7%
• Remote Disk I/O 8.5%	• Remote Disk I/O 52.7%	• Remote Disk I/O 63.4%	• Remote Disk I/O 63.0%	• Remote Disk I/O 72.3%
• Synchronization 0.1%	Network Communication 0.4%	Network Communication 0.1%	Network Communication 0.2%	Network Communication 0.7%
• Initialization 0.3%	• Synchronization 0.6%	• Synchronization 0.2%	• Synchronization 2.0%	• Synchronization 1.3%
	• Initialization 0.6%	• Initialization 0.6%	• Initialization 2.3%	• Initialization 2.3%

Figure 8-17. *Warehouse effect on MERGE*

Note the Spills to Disk value for X-Small and Small warehouses.

Having set the warehouse performance baseline, let's now segment the target core table.

Core Table Segmentation

Identifying the target core table's clustering key is crucial for developing a parallelization strategy as the target core table clustering key lead attribute is usually the prime candidate to segment the target core table. You are looking for a low-cardinality manageable range of attributes matching the most commonly used query predicates.

Good table segmentation candidates include the following:

- Geographic region

- Summary date/year range

Returning to the customer complaint, let's investigate how the application process can be improved by segmenting a core table.

While you have created sample data with known numbers of INSERT, UPDATE, and DELETE operations, you cannot parallelize using this technical dimension as the target core table is clustered using region_key and nation_key. The DML will block as the micro-partitions will be locked until each operation has completed.

Figure 8-18 illustrates the objective for this section, which is to take a single core table and segment according to the current clustering key definition.

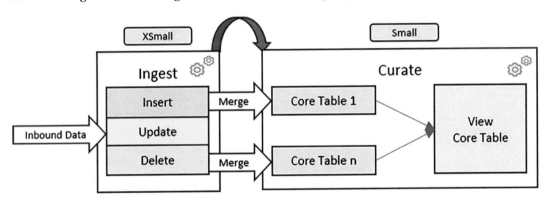

Figure 8-18. *Segmentation objective*

Before proceeding, ensure all grants to the original core table are preserved for later modification and reuse.

```
USE WAREHOUSE IDENTIFIER ( $tpc_warehouse_xs );
SHOW GRANTS ON TABLE core_customer_order;

SELECT 'GRANT '||"privilege"||' ON '||"granted_on"||' '||
       LOWER ( REPLACE ( "name", '"', '' ))||
       ' TO ROLE '||LOWER ( "grantee_name" )||';'
FROM   TABLE ( RESULT_SCAN ( last_query_id()));
```

From the previous check, you know there are five regions. You will use this information to derive the segmented tables for the target core table.

```
CREATE OR REPLACE TABLE part_customer_order_asia
(
customer_key    NUMBER(38,0),
customer_name   VARCHAR(25),
order_key       NUMBER(38,0),
order_date      DATE,
nation_name     VARCHAR(25),
region_name     VARCHAR(25)
);

CREATE OR REPLACE TABLE part_customer_order_europe
(
customer_key    NUMBER(38,0),
customer_name   VARCHAR(25),
order_key       NUMBER(38,0),
order_date      DATE,
nation_name     VARCHAR(25),
region_name     VARCHAR(25)
);

CREATE OR REPLACE TABLE part_customer_order_africa
(
customer_key    NUMBER(38,0),
customer_name   VARCHAR(25),
order_key       NUMBER(38,0),
order_date      DATE,
nation_name     VARCHAR(25),
region_name     VARCHAR(25)
);

CREATE OR REPLACE TABLE part_customer_order_america
(
customer_key    NUMBER(38,0),
customer_name   VARCHAR(25),
order_key       NUMBER(38,0),
order_date      DATE,
```

```
nation_name     VARCHAR(25),
region_name     VARCHAR(25)
);
CREATE OR REPLACE TABLE part_customer_order_middle_east
(
customer_key    NUMBER(38,0),
customer_name   VARCHAR(25),
order_key       NUMBER(38,0),
order_date      DATE,
nation_name     VARCHAR(25),
region_name     VARCHAR(25)
);
```

Now create data for each segmented table from our base table, base_customer_order.

```
USE WAREHOUSE IDENTIFIER ( $tpc_warehouse_m );

INSERT INTO part_customer_order_asia
SELECT *
FROM    base_customer_order
WHERE   region_name = 'ASIA';

INSERT INTO part_customer_order_europe
SELECT *
FROM    base_customer_order
WHERE   region_name = 'EUROPE';

INSERT INTO part_customer_order_africa
SELECT *
FROM    base_customer_order
WHERE   region_name = 'AFRICA';

INSERT INTO part_customer_order_america
SELECT *
FROM    base_customer_order
WHERE   region_name = 'AMERICA';
```

```
INSERT INTO part_customer_order_middle_east
SELECT *
FROM   base_customer_order
WHERE  region_name = 'MIDDLE EAST';
```

To preserve backward compatibility and be minimally invasive to the application, you will drop the original table and replace with a view of the same name.

```
DROP TABLE core_customer_order;

CREATE OR REPLACE VIEW core_customer_order
COPY GRANTS
AS
SELECT customer_key, customer_name, order_key, order_date,
       nation_name, region_name
FROM   part_customer_order_asia
UNION ALL
SELECT customer_key, customer_name, order_key, order_date,
       nation_name, region_name
FROM   part_customer_order_europe
UNION ALL
SELECT customer_key, customer_name, order_key, order_date,
       nation_name, region_name
FROM   part_customer_order_africa
UNION ALL
SELECT customer_key, customer_name, order_key, order_date,
       nation_name, region_name
FROM   part_customer_order_america
UNION ALL
SELECT customer_key, customer_name, order_key, order_date,
       nation_name, region_name
FROM   part_customer_order_middle_east;
```

Confirm the row count from the new view, `core_customer_order`, matches the original table row count of 1,500,000,000.

```
USE WAREHOUSE IDENTIFIER ( $tpc_warehouse_xs );

SELECT count(1)
FROM   core_customer_order;
```

To preserve backward compatibility, re-create the grants for the original table, which would look something like this, substituting in the role names to match your environment:

```
GRANT INSERT, UPDATE, DELETE, TRUNCATE ON TABLE
part_customer_order_asia       TO ROLE <YOUR_INGEST_ROLE>;
GRANT INSERT, UPDATE, DELETE, TRUNCATE ON TABLE
part_customer_order_europe     TO ROLE <YOUR_INGEST_ROLE>;
GRANT INSERT, UPDATE, DELETE, TRUNCATE ON TABLE
part_customer_order_africa     TO ROLE <YOUR_INGEST_ROLE>;
GRANT INSERT, UPDATE, DELETE, TRUNCATE ON TABLE
part_customer_order_america    TO ROLE <YOUR_INGEST_ROLE>;
GRANT INSERT, UPDATE, DELETE, TRUNCATE ON TABLE
part_customer_order_middle_east TO ROLE <YOUR_INGEST_ROLE>;

GRANT SELECT ON VIEW
core_customer_order TO ROLE <YOUR_CONSUMER_ROLE>;
```

With the target core table segmented and reconstituted via a view, let's now investigate parallelizing data ingestion.

Concurrent Warehouse Processing

Each segment requires a matching warehouse processing unit as you will shard the data to match the clustering key range.

Figure 8-19 shows how you can automate the data ingest using one pair of streams and tasks per partition; you also use a common stored procedure. Using a common stored procedure enables a parameter-driven approach to parallelization. Alternative orchestration tooling exists, which I will leave to you for your further investigation.

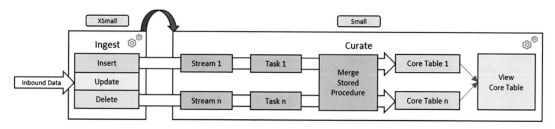

Figure 8-19. *Concurrent processing pattern*

This example uses five partition tables, which may reduce the warehouse size from Medium to Small. As recommended throughout this book, testing will determine the optimal warehouse size, and any suggestion on appropriate warehouse size must be proven.

With five partition tables, you will require five streams, five tasks, and a single parameterized stored procedure.

You must create streams before populating raw or staging tables with data to ensure each stream registers the loaded data.

DML order is significant: truncate *before* creating streams.

To ensure the steams are populated, you first use TRUNCATE on the staging table stg_customer_order and then reload each data set for INSERT, UPDATE, and DELETE.

```
TRUNCATE TABLE stg_customer_order;
```

Create five streams on the raw or staging table, one for each target partition table.

```
CREATE OR REPLACE STREAM strm_part_customer_order_asia
ON TABLE stg_customer_order;
CREATE OR REPLACE STREAM strm_part_customer_order_europe
ON TABLE stg_customer_order;
CREATE OR REPLACE STREAM strm_part_customer_order_africa
ON TABLE stg_customer_order;
CREATE OR REPLACE STREAM strm_part_customer_order_america
ON TABLE stg_customer_order;
CREATE OR REPLACE STREAM strm_part_customer_order_middle_east
ON TABLE stg_customer_order;
```

Now populate the raw or staging table `stg_customer_order` using the known base data for each DML operation from earlier.

```
USE WAREHOUSE IDENTIFIER ( $tpc_warehouse_m );
INSERT OVERWRITE INTO stg_customer_order
SELECT *
FROM    base_customer_order_insert
UNION
SELECT *
FROM    base_customer_order_update
UNION
SELECT *
FROM    base_customer_order_delete;
```

With the test data configured and the target core table reconfigured as five segmented tables, you can do the following:

```
USE WAREHOUSE IDENTIFIER ( $tpc_warehouse_xs );


SELECT count(1)
FROM    part_customer_order_asia;
```

For this data set, you see 300,094,996 records; your record count will differ. Make a note of this number.

```
SELECT count(1)
FROM    stg_customer_order
WHERE   region_name = 'ASIA';
```

For this data set, you see 60,417 records; your record count will differ. Make a note of this number.

Stream Interaction

Streams are an ideal mechanism for identifying change data capture on a base object, but their usage can be confusing. Let's examine use cases for integrating streams into our data pipeline.

Testing Streams

When developing a test case, you carry out these steps:

1. Create a staging table.

2. Populate the staging table with test data.

3. Run a MERGE statement.

4. TRUNCATE the staging table.

5. Create five streams.

6. Populate the staging table with test data.

If you examine the contents of a single stream, you observe a single record (step 6).

```
SELECT count(1),
       metadata$action,
       metadata$isupdate
FROM   strm_part_customer_order_asia
WHERE  region_name = 'ASIA'
GROUP BY metadata$action,
         metadata$isupdate;
```

You should see 30,000 INSERTs, as shown in Figure 8-20.

COUNT(1)	METADATA$ACTION	METADATA$ISUPDATE
300000	INSERT	FALSE

Figure 8-20. *Stream output*

By design intent streams do not record UPDATEs but instead record UPDATEs as a pair of INSERT and DELETE operations where metadata$isupdate is set to true. Where metadata$isupdate is set to false, this indicates the DML operations are not related. Use metadata$row_id to correlate INSERT and DELETE pairs, an exercise left to you.

For this use case you can use the presence of data in a STREAM to trigger a TASK, which calls a stored procedure where you consume from the stream.

Streams can go stale where the data is not consumed within the retention period and ensure each stream is cleared out before reuse. You can find more information on streams going stale at `https://community.snowflake.com/s/article/The-query-that-reads-or-consumes-the-stream-is-failing`. When a stream goes stale, re-creating the stream will solve the issue though the contained data may be lost.

In this section I have called out how our specific implementation uses streams; you can learn more about steams at `https://docs.snowflake.com/en/sql-reference/sql/create-stream`.

Creating Stored Procedures

Let's create a stored procedure passing `region_name` as a parameter that must match the segment suffix you are processing.

```
CREATE OR REPLACE PROCEDURE sp_merge_test_load ( P_REGION STRING )
RETURNS string
LANGUAGE javascript
EXECUTE AS CALLER
AS
$$
   var sql_stmt  = "";
   var err_state = "";
   var result    = "";

   sql_stmt  = "MERGE INTO part_customer_order_" + P_REGION + " c\n"
   sql_stmt += "USING strm_part_customer_order_" + P_REGION + " s\n"
   sql_stmt += "ON    c.order_key   = s.order_key\n"
   sql_stmt += "AND   s.region_name = '" + P_REGION + "'\n"
   sql_stmt += "WHEN MATCHED AND s.operation = 'D' THEN DELETE\n"
   sql_stmt += "WHEN MATCHED AND s.operation = 'U' THEN\n"
   sql_stmt += "   UPDATE SET customer_key  = s.customer_key,\n"
   sql_stmt += "              customer_name = s.customer_name,\n"
   sql_stmt += "              order_date    = s.order_date,\n"
   sql_stmt += "              nation_name   = s.nation_name,\n"
   sql_stmt += "              region_name   = s.region_name\n"
   sql_stmt += "WHEN NOT MATCHED\n"
```

```
   sql_stmt += "    AND s.operation = 'I'\n"
   sql_stmt += "    AND s.region_name = '" + P_REGION + "' THEN\n"
   sql_stmt += "        INSERT ( customer_key,\n"
   sql_stmt += "                 customer_name,\n"
   sql_stmt += "                 order_key,\n"
   sql_stmt += "                 order_date,\n"
   sql_stmt += "                 nation_name,\n"
   sql_stmt += "                 region_name )\n"
   sql_stmt += "        VALUES ( s.customer_key,\n"
   sql_stmt += "                 s.customer_name,\n"
   sql_stmt += "                 s.order_key,\n"
   sql_stmt += "                 s.order_date,\n"
   sql_stmt += "                 s.nation_name,\n"
   sql_stmt += "                 s.region_name );\n"

   stmt = snowflake.createStatement( { sqlText: sql_stmt } );
   try
   {
      stmt.execute();
      result = "Success: Rows Affected: " + stmt.getNumRowsAffected()
      + " Deleted: " + stmt.getNumRowsDeleted() + " Updated: " + stmt.
      getNumRowsUpdated() + " Inserted: " + stmt.getNumRowsInserted();
   }
   catch(err)
   {
      err_state += "\nFail Code: " + err.code;
      err_state += "\nState: " + err.state;
      err_state += "\nMessage: " + err.message;
      err_state += "\nStack Trace:" + err.StackTraceTxt;
      err_state += "\nSQL Statement:\n\n" + sql_stmt;
      result = err_state;
   }
   return result;
$$;
```

With the stored procedure created and an understanding of how streams operate, let's now test.

Testing a Single Load

Now test a single load:

```
CALL sp_merge_test_load ( 'ASIA' );
```

The stored procedure should return a message similar to this:

Success: Rows Affected: 60417 Deleted: 20304 Updated: 20067 Inserted: 20046

Confirm the stream contents have been consumed.

```
SELECT count(1),
       metadata$action,
       metadata$isupdate
FROM   strm_part_customer_order_asia
WHERE  region_name = 'ASIA'
GROUP BY metadata$action,
         metadata$isupdate;
```

Check the number of records in our table partition.

```
SELECT count(1)
FROM   part_customer_order_asia; //300094738
```

Confirm you have the correct results from the MERGE stored procedure. The values are the before and after row counts from the table partition.

```
SELECT 300094738 - 300094996; //-258
```

Using the stored procedure return values, subtract the DELETED row count from the INSERT row count. This number represents the net difference. You can ignore the UPDATE row counts as these do not change the number of rows in the table partition.

```
SELECT 20046     - 20304;       //-258
```

While the stored procedure's before and after checks work for my tests, your values will differ because of the use of LIMIT when generating test data.

Grant Entitlement

Grant entitlement to role tpc_owner_role to manage tasks.

```
USE ROLE securityadmin;

GRANT CREATE TASK ON SCHEMA tpc.tpc_owner TO ROLE tpc_owner_role;

USE ROLE accountadmin;

GRANT EXECUTE TASK ON ACCOUNT TO ROLE tpc_owner_role;
```

Reset the role to tpc_owner_role.

```
USE ROLE IDENTIFIER ( $tpc_owner_role );
```

Create Tasks

With the stored procedure and stream integration proven to work correctly, you can now automate data ingestion by creating five tasks, one for each table segment.

Alternative methods of scheduling are available:

- Tasks are advantageous as they exist within the confines of Snowflake. There are no external dependencies, but they incur latency because of the scheduled trigger timer.

- External scheduling tools are advantageous for orchestrating sequential data load and stored procedure execution without incurring time delays between processing steps.

For testing purposes only, the SCHEDULE is set to 1 minute. In your real-world scenario, the SCHEDULE should be set to a more representative value according to expected raw or staging data arrival time.

```
CREATE OR REPLACE TASK tsk_part_customer_order_asia
WAREHOUSE = tpc_wh_small
SCHEDULE  = '1 minute'
WHEN system$stream_has_data ( 'strm_part_customer_order_asia' )
AS
CALL sp_load_customer( 'ASIA' );
```

```
CREATE OR REPLACE TASK tsk_part_customer_order_europe
WAREHOUSE = tpc_wh_small
SCHEDULE  = '1 minute'
WHEN system$stream_has_data ( 'strm_part_customer_order_europe' )
AS
CALL sp_load_customer( 'EUROPE' );

CREATE OR REPLACE TASK tsk_part_customer_order_africa
WAREHOUSE = tpc_wh_small
SCHEDULE  = '1 minute'
WHEN system$stream_has_data ( 'strm_part_customer_order_africa' )
AS
CALL sp_load_customer( 'AFRICA' );

CREATE OR REPLACE TASK tsk_part_customer_order_america
WAREHOUSE = tpc_wh_small
SCHEDULE  = '1 minute'
WHEN system$stream_has_data ( 'strm_part_customer_order_america' )
AS
CALL sp_load_customer( 'AMERICA' );

CREATE OR REPLACE TASK tsk_part_customer_order_middle_east
WAREHOUSE = tpc_wh_small
SCHEDULE  = '1 minute'
WHEN system$stream_has_data ( 'strm_part_customer_order_middle_east' )
AS
CALL sp_load_customer( 'MIDDLE_EAST' );
```

Now enable each task.

```
ALTER TASK tsk_part_customer_order_asia          RESUME;
ALTER TASK tsk_part_customer_order_europe        RESUME;
ALTER TASK tsk_part_customer_order_africa        RESUME;
ALTER TASK tsk_part_customer_order_america       RESUME;
ALTER TASK tsk_part_customer_order_middle_east RESUME;
```

Prove all tasks are scheduled.

```
SELECT timestampdiff ( second, current_timestamp, scheduled_time ) AS
next_run,
        scheduled_time,
        current_timestamp,
        name,
        state
FROM TABLE ( information_schema.task_history())
ORDER BY completed_time DESC;
```

Figure 8-21 shows example output for scheduled tasks.

NEXT_RUN	SCHEDULED_TIME	CURRENT_TIMESTAMP	NAME	STATE
56	2024-01-01 07:38:58.448 -0800	2024-01-01 07:38:02.675 -0800	TSK_PART_CUSTOMER_ORDER_MIDDLE_EAST	SCHEDULED
56	2024-01-01 07:38:58.170 -0800	2024-01-01 07:38:02.675 -0800	TSK_PART_CUSTOMER_ORDER_AMERICA	SCHEDULED
55	2024-01-01 07:38:57.918 -0800	2024-01-01 07:38:02.675 -0800	TSK_PART_CUSTOMER_ORDER_AFRICA	SCHEDULED
55	2024-01-01 07:38:57.590 -0800	2024-01-01 07:38:02.675 -0800	TSK_PART_CUSTOMER_ORDER_EUROPE	SCHEDULED
55	2024-01-01 07:38:57.222 -0800	2024-01-01 07:38:02.675 -0800	TSK_PART_CUSTOMER_ORDER_ASIA	SCHEDULED

Figure 8-21. *Stream output*

Purging a Stream

After successfully merging all raw or staged data, you must remove the data in preparation for the next load. You may perform the TRUNCATE just before loading new data; preserving raw or staged data until just prior to the next load is good practice in the event you need to investigate the most recent data load. However, TRUNCATE has a side effect of also removing load metadata; you can find information at https://docs.snowflake.com/en/sql-reference/sql/truncate-table#usage-notes.

In this example you use TRUNCATE, which is not the only option for clearing a staging table. You can choose INSERT OVERWRITE instead. Regardless of the method chosen, interaction with our choice of data load operator must be tested to ensure the stream correctly expresses the desired outcome.

```
TRUNCATE TABLE stg_customer_order;
```

For further information, please learn more about the COPY command: https://docs.snowflake.com/en/sql-reference/sql/copy-into-table?utm_source=legacy&utm_medium=serp&utm_term=copy and for Snowpipe: https://docs.snowflake.com/en/user-guide/data-load-snowpipe-intro?utm_source=legacy&utm_medium=serp&utm_term=snowpipe.

Let's now check what the stream recorded.

```
SELECT count(1),
       metadata$action,
       metadata$isupdate
FROM   strm_part_customer_order_asia
WHERE  region_name = 'ASIA'
GROUP BY metadata$action,
         metadata$isupdate;
```

However, you find our stream registers a DELETE operation for all staged data, as shown in Figure 8-22.

COUNT(1)	METADATA$ACTION	METADATA$ISUPDATE
60417	DELETE	FALSE

Figure 8-22. *Stream TRUNCATE data*

You must purge the stream DELETE data before loading our next batch into the raw or staging table. Snowflake does not provide capability to purge stream contents, but you know a simple SELECT will *not* clear the stream content.

To clear stream contents you must SELECT all records using a dummy INSERT as this next statement proves:

```
INSERT INTO part_customer_order_asia
SELECT *
FROM   strm_part_customer_order_asia
WHERE  1 = 0;
```

Now recheck the stream contents.

```
SELECT count(1),
       metadata$action,
       metadata$isupdate
FROM   strm_part_customer_order_asia
WHERE  region_name = 'ASIA'
GROUP BY metadata$action,
         metadata$isupdate;
```

You should see zero records.

Suspend Tasks

After testing, suspend tasks to prevent inadvertent execution.

```
ALTER TASK tsk_part_customer_order_asia        SUSPEND;
ALTER TASK tsk_part_customer_order_europe       SUSPEND;
ALTER TASK tsk_part_customer_order_africa       SUSPEND;
ALTER TASK tsk_part_customer_order_america      SUSPEND;
ALTER TASK tsk_part_customer_order_middle_east SUSPEND;
```

Load Testing

Load testing can be conducted in several ways assuming the raw or staging table has been pre-populated.

- Amend the External Parallelism Component developed in Chapter 6 to call the stored procedure five times, one call for each region.

- Resume all five tasks.

- Use an external orchestration tool.

Parallelizing data pipelines is dependent upon matching unused processing units in a warehouse to the number of concurrent processes required to partition the underlying table. You assume eight processing units; therefore, you should aim to split a single table into eight segments. As you experienced with segmenting by region, there are only five regions and hence five segments.

Assuming all parallel operations occur simultaneously, you expect to see both an overall reduction in execution time and a higher concurrent use of a single warehouse. Note the use of a task and stream may later be replaced by a single dynamic table currently in public preview.

I suggest repeating the load test using a smaller warehouse size while checking for both queueing and spills to disk to optimize cost and performance. The expectation is that parallel processes will perform well with smaller warehouses.

I covered parallel testing in Chapter 6 and therefore leave testing for your further investigation.

Concluding Steps

In concluding the test case, you carry out these steps:

1. Create a stored procedure to load segments.

2. Test a single data load.

3. Create and run tasks and then suspend tasks.

4. Clean up to get ready for the next load.

While the number of segments may vary along with the orchestration tooling, the technique is sound and delivers measurable performance benefits in real-world use.

Temporal Loads

This test case considered a fictitious scenario that you developed into parallelizing a single load into five separate region segments. The feed may not conform to the same pattern, and furthermore, our data may skew over time.

Let's assume you have a temporal feed where the bulk of data changes over time. Imagine a feed where the majority of the data is for a sliding three-year window. The segmentation strategy must adapt to cater to the feed, so the steps involved include the following:

- Identify the date key and create segments for date ranges.

- Set the segment date ranges as follows:

  - Large for low-volume changes

  - Small for high-volume changes

- Implement annual segment maintenance for the feed sliding window content.

Figure 8-23 illustrates a sample date range with relative data volumes.

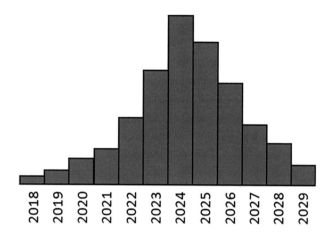

Figure 8-23. *Temporal load sample data*

From the information presented in Figure 8-23, you can deduce the following:

- The record date ranges are increasing through time with more future dated records appearing and fewer older dated records.

- The highest volume of new, changed, or deleted records occurs for 2023, 2024, 2025, and 2026.

- The data is skewed; i.e., the records are not evenly distributed according to date.

This segmentation approach must reflect the sample date ranges; therefore, our segment ranges might look similar to Figure 8-24.

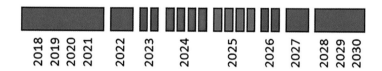

Figure 8-24. *Temporal load partitions*

Not all partitions will contain equal date ranges to reflect the relative volume of DML operations for the period. Note the focus on 2024 and 2025 where I suggest four partitions matching each quarter year.

As the load data profile moves forward in time, you must periodically adjust the date ranges for the partitions and remap the data loads to match. I suggest the maintenance is conducted annually as an end of year activity.

While the sample data does not contain records for 2030, it is prudent to extend the highest date range partition into the future to capture any outliers. For the same reason, the earliest date range partition should start well before 2018.

The sharp-eyed reader will note the suggested number of partitions matches a two-cluster X-Small warehouse. Where the opportunity exists, you should maximize warehouse processor unit consumption and favor multiples of eight when partitioning tables.

Real-World Impact

The practical application of the parallelization technique outlined in this chapter in a real-world production environment delivers significant business benefit. When applied selectively, you may see between 40 percent and 70 percent reduction in query response times in like-for-like scenarios along with a 25 percent reduction in ingestion and curation time leading to more rapid data product delivery to clients.

You also experienced an unexpected benefit: parallelizing processing may reduce the number of micro-partitions churned. When replicating tables to other accounts, the cost of replication can exceed the cost of data product curation. Reducing micro-partition churn significantly reduces replication costs.

Improving overall system performance, throughput, and cost reduction using this technique may expose a hidden issue: the increase in concurrent processing can lead to an increase in the volume of logging into a single table. The inadvertent serialization of logging causes queueing as each process locks the log table. The solution is to implement EVENT-based logging as described in Chapter 6. This is the Law of Unintended Consequences in action again.

I must point out the selective adoption of this technique; I strongly suggest this technique is *not* implemented ubiquitously. There are no silver bullets, and this technique is no exception. If poorly implemented, this technique may experience higher overall warehouse cost when parallelizing our data pipelines. However, increased data curation costs may be offset by a reduction in runtimes. Every change you make requires testing, and implementing parallelization is no different.

Summary

This chapter by setting the scene for wider data product distribution and explored an example application profile calling out the typical sections along with capabilities. I assigned nominal warehouse sizes to each section as a starting point.

After establishing an example application profile, I covered a typical end-user complaint and provided information and tools to analyze the root cause, capturing useful information along the way. You investigated how to segregate a data load by DML operation and learned why you cannot run these three operations concurrently.

You then investigated how to parallelize an existing data feed by creating segmented data sets according to a known business key derived from a core table clustering key. This work led to the creation of five separate automated data pipelines along with test cases for each step along the way. I called out some side effects of DML operations relating to streams along with mitigating actions to support rerun capability.

Temporal data loads were also discussed to expose an alternative date-based segmentation strategy along with identifying the annual maintenance overhead.

I concluded by assessing the real-world impact of implementing parallelization, noting the technique outlined in this chapter is not a silver bullet and should not be applied ubiquitously.

Having discussed parallelization in depth, we will now move on to investigate entitlements.

CHAPTER 9

Client Expectations

This chapter covers how to tune your approach to client interactions. Reducing both cost and time for your client is a key selling point of your products. Your client expectations are critically important in delivering successful business outcomes.

Curated data products are the result of applying your organization's intellectual property to data to realize a commercial offering. I discussed the DIKW pyramid in Chapter 8, showing the relationship between data, information, knowledge, and wisdom; for more information, see https://en.wikipedia.org/wiki/DIKW_pyramid.

A simple example of a curated data product is extracting data marketplace revenue figures from financial reports and showing trends over time. The intellectual property could include identifying and collating the raw data from differently formatted company reports, applying your bespoke logic, and then presenting information in a simple manner showing the historical trends. You might enrich your report with relevant supporting context such as links to each company website and then deliver your report as part of a comprehensive market analysis to clients.

Many legacy source data sets are currently points in time only; that is, only the current view of data is available, and the latest changes overwrite the current records. Later in this chapter, I will discuss how to provide historical point-in-time reporting. Many clients want the ability to re-create reports for any given time period. By utilizing Snowflake's built-in capability, you can serve up temporal data to provide this additional commercial opportunity.

This chapter focuses on how you can deliver your curated data products to your clients, preferably exceeding their expectations. Presenting a consistent, well-articulated approach supported by a trusting relationship often results in increased sales. Also, a happy customer consumes more and demands less from your support functions and can help your organization through positive feedback and critiques.

In support of a "go to market" proposal, a data distribution strategy should address multi-platform data interchange and cross-platform data sourcing for augmentation to which Snowflake is a significant contributor. Your client experience may be wider

© Andrew Carruthers 2024
A. Carruthers, *Tuning the Snowflake Data Cloud*, https://doi.org/10.1007/979-8-8688-0379-6_9

than consuming from Snowflake, something to be kept in mind when addressing client expectations. Regardless of how clients source your data products, you must deliver a consistent experience across all platforms.

Let's look at the client perspective: your clients want highly performant data products delivered in their specified consumable formats to their operating locations within agreed timeframes. Increasingly, your clients are becoming more aware of their dependence upon your ability to serve data products in a resilient manner. No consumer should be forced into invoking their disaster recovery process as a consequence of provider infrastructure failure. You must insulate your consumers as much as possible.

You must consider that your data product offerings are one or more data source ingestion boxes within your client's architecture diagrams. In other words, your data products may not be central to your client's business; there are plenty of competitors out there, and your approach must align with your clients' requirements.

Companies do not build data products speculatively. The available evidence proves that "if you build it, they will *not* come." Internal organization data product consumption is a side benefit. While notable exceptions exist (such as COVID-19 data), the typical purpose for building data products is for generating revenue for your organization.

When viewed from a client perspective, you must deliver against their requirements and consider yourself a valued contributor to their success. Your value-add must include data dictionaries, catalogs, and entity-relationship diagrams to inform clients of entity relationships, business keys, and technical keys. Furthermore, if your organization provides multiple data products, you should demonstrate where the data models intersect as this may lead to up-sell opportunities.

When provisioning shares, there is no additional cost to adding multiple accounts within the same CSP and region. For example, let's assume a client has three Snowflake accounts on the same CSP and region, one each for development, testing, and production. Enabling the same share for consumption by all three accounts is a simple operation; you would enable all three accounts to reference the same read-only data in a consistent manner. Sometimes clients want to use the same data for testing as they would in production. No more copying purchased data sets from a single share to multiple accounts. I do acknowledge there may be usage license implications for the additional service provision. Making the same shares available removes the need to copy data, reduces your client costs, and provides an up-sell opportunity for additional user licenses.

While your primary focus is to both provision and entitle your curated data products to enable consumption by your clients, you must do so in ways that both reduce consumption friction and keep costs as low as possible. These are some examples:

- You can insulate your clients from internal delivery failure.

- Producing client-specific prefiltered data prevents navigation of an entitlement model for each SQL call.

- Delivering a data catalog describing relationships and interactions enables rapid data product integration with client data.

- Supplying sample SQL statements provides real-world examples to leverage data products.

All of this reduces the total cost of ownership (TCO), improves system performance, and removes barriers to adoption for your clients.

Previous chapters focused on technical details supporting application performance and curation of data products; this chapter focuses on how your clients gain access and interact with data products along with provisioning an extended suite of tooling. Your goal is to provision your data products along with contextual information to enable your clients to rapidly understand, assimilate, and integrate into their environments.

Later in this chapter I will discuss the wider implications of delivering data products into disparate marketplaces, providing a wider context for your further investigation.

Let's start with discussing how you entitle your data products.

Entitlement Models

Regardless of the distribution venue for your data products, implementing entitlement models to ensure a client accesses their licensed data properly incurs both cost and time.

Your approach must work for direct access to the local account where your data products are curated, for access via an imported Secure Direct Data Share, or for access via a replicated database.

In this chapter I discuss two entitlement model approaches.

- Embedded into client-accessible objects

- Pre-filtered, client-specific data objects

Both entitlement models have their pros and cons. Figure 9-1 shows both entitlement models side by side.

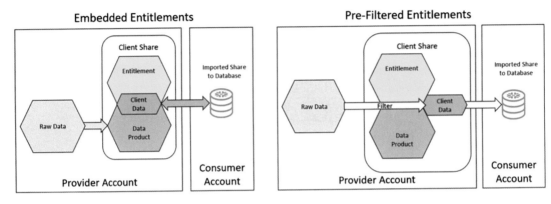

Figure 9-1. *Entitlement models*

We unpack both entitlement models next; note that both models use the same target objects albeit with curated data sets.

Embedded Entitlement Model

Most entitlement models are "baked in" to the end user's queried objects. The reason is simple: it's an easy way for developers to deliver quickly at acceptable performance levels. However, over time, performance often degrades as the data sets become larger or skewed and the entitlement model becomes more complex. Embedded entitlement models cost more to maintain over time.

Embedded entitlement models are typically in the form of SQL predicates joining entitlement objects to data product objects. Each SQL invocation (unless results are cached) results in the re-evaluation of the entitlement model to identify and return appropriate data. Figure 9-2 illustrates a typical overview of an embedded entitlement model.

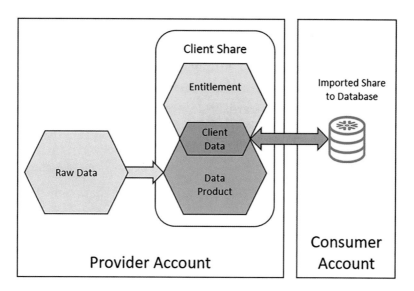

Figure 9-2. *Embedded entitlement model*

There are alternative entitlement options including API calls that may be suitable for low-volume data sets, but I do not discuss these further.

Embedded entitlement models have some advantages.

- They are easy to implement; one size fits all.

- Shares contain identical copies of objects containing identical data.

- Provisioning is simple.

- Data is available at the point of curation.

However, there are some disadvantages.

- Every SQL query with the exclusion of cached result sets navigates the entitlement model to derive result sets.

- Individual query performance issues can be hard to solve.

- Changes to the entitlement model can be pervasive affecting many objects and requiring extensive retesting.

- Replicating all data can be costly, particularly where only subsets of data are accessed by clients.

Embedded entitlement models are sometimes encountered in legacy systems ported to Snowflake. You should also be aware ported code may not be optimized for Snowflake.

Embedded entitlement models can be difficult to understand and may contain bespoke rules within query predicates. Sometimes, in an attempt to abstract entitlement, several layers of object may contain partial rule sets; beware of views calling views!

Prefiltered Entitlement

An alternative approach is to prefilter data to present entitled data only. Figure 9-3 illustrates a typical overview of a prefiltered entitlement model.

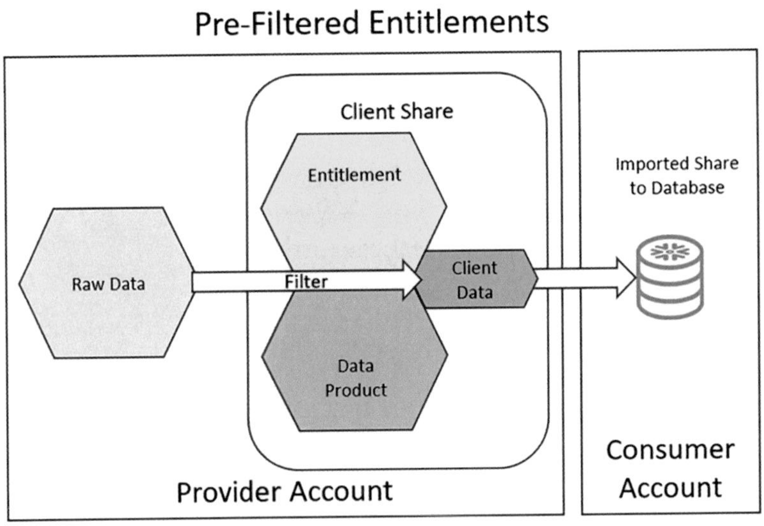

Figure 9-3. *Prefiltered entitlement model*

Prefiltered entitlement models have some advantages.

- The smaller data sets are curated for each client.

- Performance issues are more easily resolved.

- SQL queries do not navigate an entitlement model.

- The changes to the entitlement model are localized to the filter engine.

- Only the client-consumed data is replicated.

However, there are some disadvantages.

- They are more complex to implement; each data set is bespoke.

- Provisioning is more complex.

- There is a proliferation of source objects; the objects are the same, but the content differs.

- Data is not immediately available at the point of curation.

Prefiltered entitlement models are not common; complex implementation is often discounted for the benefit of a simple but less performant embedded SQL implementation. The additional effort required to develop a pre-filtered entitlement model will deliver significant benefits as both consumption grows and your data products mature.

Having identified the two patterns for delivering entitled data products, you now know how to build a filter engine to support the bespoke curation of data products for each individual client.

Filter Engine Overview

Figure 9-4 provides an overview of the entities required for a filter engine.

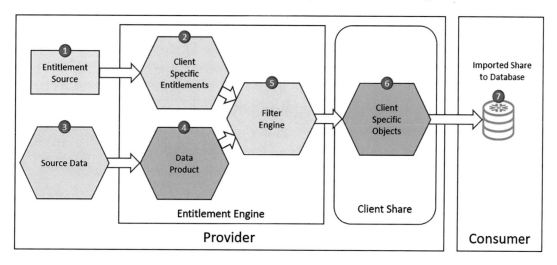

Figure 9-4. *Filter engine overview*

The functional components shown within Figure 9-4 are as follows:

1. Entitlements provided by an external entitlement component.

2. A normalized entitlement data model containing

   a. Client-specific entitlement

   b. Mapping to data product objects

   c. Template SQL statements

3. Various source data feeds

4. Curated data product objects

5. Filter engine that

   a. Maps entitlement to data product object

   b. Applies filters to generate client-specific content

6. Client-specific share containing entitled data

7. Imported share manifests as a database in the consumer account

Let's investigate each component in more detail.

External Entitlement Component

Many organizations experience growth through a merger and acquisition (M&A), which results in the proliferation of entitlement applications: every acquired product has its own entitlement system. Plugging the gaps in data product offerings via the M&A activity involves acquiring a corresponding entitlement application.

Without a single strategic entitlement application, acquired data products cannot be fully integrated with existing data products for entitlement purposes.

I am assuming entitlements are sourced and then merged into your normalized entitlement data model.

Entitlement Data Model

The absence of a single strategic entitlement application implies you may require more than one entitlement data model—one for each source entitlement system. For the purposes of developing your filter engine, you assume a single entitlement source.

Figure 9-5 shows the key components of an entitlement data model and usage. Note that The filter and client objects are shown for context only.

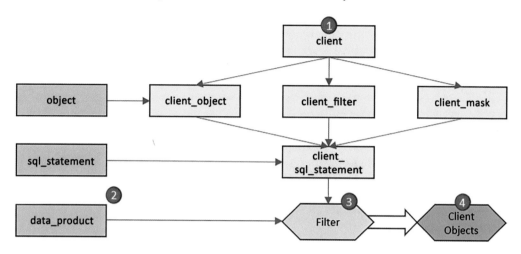

Figure 9-5. *Entitlement data model*

The entitlement model must contain the following:

- Client information mapped to objects, filters, masks, and template SQL statements

- Data product objects, reference data, and template SQL statements

Source Data Feeds

Source data feeds are not discussed here as they should be self-explanatory for consumption and use.

Curated Data Product

Curated data products are the unentitled superset of generated products sold to your customers; these data products combine various data sources with your intellectual property. Data products are constantly maintained as new data becomes available; the curation process is constantly ongoing.

For internal use cases, you may choose to distribute unentitled data to trusted internal consumers where their local entitlement overlays may apply. This distinction is important: internal data product distribution use cases are much simpler to implement and easier to gain approval for than external data product distribution use cases.

Filter Engine

The absence of client object mapping precludes the use of SQL statements for generating client-specific objects. SQL statements act as overlays to the underlying data product objects that incorporate all client-specific filters and masks.

The previous statement is a bit of a mouthful to say and sounds complicated, and in truth, generating bespoke objects is an advanced topic. I demonstrate how to do this later.

Filter engine output should be thoroughly tested before deployment. Only when confident should you consider automated deployment, along with corresponding generated test cases.

Client-Specific Shares

Snowflake shares are structural container objects created by the ACCOUNTADMIN role. Share ownership may be transferred to other roles. Note that a single role can hold this privilege on only a specific share object at a time. Semi-automated client-specific sharing relies upon the dynamic generation of a share and schema, a Data Definition Language (DDL) entitlement, and finally assigning the share to a nominated Snowflake account.

When the dynamic requirements have been satisfied, you can deploy the filter engine output; note that the data content for your objects should differ for each client. A hybrid approach is to generate a common suite of unentitled objects for bulk distribution alongside a bespoke defined suite of components. Your use cases will inform your decision.

Once a share has been authorized to an account, importing the share appears as a database within the Snowflake user interface. The imported database requires local client administration to make the generated data products accessible.

Unentitled Data Sharing

To remove the need to create a second Snowflake account within the same CSP and region, you begin by creating a managed account. You then continue by creating the containers to deliver unentitled objects and a share to your fictitious customer, after which you expand your delivery for entitled objects to the same fictitious customer.

Let's get started!

Creating Managed Accounts

Managed accounts (also known as reader accounts) exist within the context of a single Snowflake account. According to the Snowflake documentation:

> "...Enable data consumers to access and query data shared by the provider of the account, with no setup or usage costs for the consumer, and no requirements for the consumer to sign a licensing agreement with Snowflake."

When sharing a "share" with a second Snowflake account, this section is not required.

Managed accounts use the *same* credit allocation as the primary account. Warehouses in the managed account can consume unlimited credits; therefore, a resource monitor to limit usage should be set up. You can find more information on resource monitors at `https://docs.snowflake.com/en/sql-reference/sql/create-resource-monitor`.

Figure 9-6 shows the relationship between Snowflake-supported containers used to implement data sharing and the tight coupling between a primary account and a managed account.

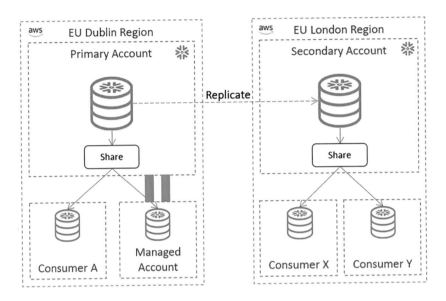

Figure 9-6. *Managed account in context*

Let's now create a managed account for which you require the ACCOUNTADMIN role.

```
USE ROLE accountadmin;
```

Password restrictions apply, including minimum length and case sensitivity.

```
CREATE MANAGED ACCOUNT poc
ADMIN_NAME      = 'poc_admin'
ADMIN_PASSWORD = 'POC_admin_123'
TYPE            = READER
COMMENT         = 'POC Managed Account';
```

We should see a JSON string returned containing your managed account information.

```
{
  "accountName": "POC",
  "accountLocator": "HR83528",
  "url": "https://acxelcq-poc.snowflakecomputing.com",
  "accountLocatorUrl": "https://hr83528.eu-west-2.aws.
  snowflakecomputing.com"
}
```

Similar information is available using the SHOW command.

```
SHOW MANAGED ACCOUNTS;
```

Please make a note of the JSON accountLocator or locator attribute, which you will use to import the share. In this example, this value is HR83528.

You will also require the accountLocatorURL or account_locator_url value to create a user, as shown next. In this example, the value is https://hr83528.eu-west-2. aws.snowflakecomputing.com/.

You can find more information on managed accounts at https://docs.snowflake. com/en/user-guide/data-sharing-reader-create.

Creating Share Containers

You will extend your tpc database by creating new containers. Figure 9-7 shows both existing and new containers you will create within this section.

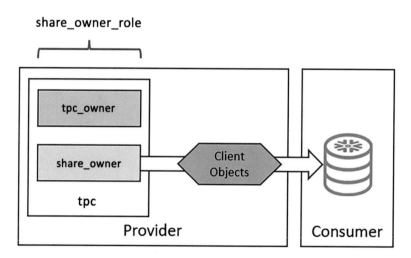

Figure 9-7. *Container creation*

First create a share called share_poc where poc represents "proof of concept":

```
USE ROLE accountadmin;

CREATE SHARE IF NOT EXISTS share_poc;
```

To display all shares within your account, use this:

```
SHOW SHARES IN ACCOUNT;
```

You can find more information on shares at https://docs.snowflake.com/en/ user-guide/data-sharing-provider#preparing-to-create-a-share.

Now create a new schema called tpc.share_owner to contain client-specific objects for assignment to share_poc. In your real-world implementation, you may want to rename share_poc to contain client-specific identifiers for ease of later identification.

```
USE ROLE sysadmin;

CREATE SCHEMA IF NOT EXISTS tpc.share_owner;
```

Create a new role to manage shared objects and assign them to yourself:

```
USE ROLE securityadmin;

CREATE OR REPLACE ROLE share_owner_role;

GRANT ROLE share_owner_role TO USER <YOUR USER HERE>;
```

Grant entitlement for the share to reference objects within the new schema called tpc.share_owner created earlier.

```
GRANT USAGE        ON DATABASE tpc           TO SHARE share_poc;
GRANT USAGE        ON SCHEMA   tpc.share_owner TO SHARE share_poc;
```

Grant entitlement to the new role called share_owner_role for creating objects in the new schema tpc.share_owner.

```
GRANT USAGE        ON DATABASE tpc           TO ROLE share_owner_role;
GRANT USAGE        ON SCHEMA   tpc.tpc_owner   TO ROLE share_owner_role;
GRANT USAGE        ON SCHEMA   tpc.share_owner TO ROLE share_owner_role;
```

Grant entitlement to the new role share_owner_role for accessing objects in both existing schema the tpc.tpc_owner and the new schema tpc.share_owner.

```
GRANT SELECT ON ALL TABLES          IN SCHEMA tpc.tpc_owner   TO ROLE
                                    share_owner_role;
GRANT SELECT ON ALL DYNAMIC TABLES  IN SCHEMA tpc.tpc_owner   TO ROLE
                                    share_owner_role;
GRANT SELECT ON ALL TABLES          IN SCHEMA tpc.share_owner TO ROLE
                                    share_owner_role;
```

```
GRANT SELECT ON ALL DYNAMIC TABLES      IN SCHEMA tpc.share_owner TO ROLE
                                        share_owner_role;
GRANT SELECT ON ALL MATERIALIZED VIEWS IN SCHEMA tpc.share_owner TO ROLE
                                        share_owner_role;
```

Snowflake security model does not allow the creation of GRANTs for objects created in the future. Attempting to do so generates an error, for example:

```
GRANT SELECT ON FUTURE TABLES           IN SCHEMA tpc.share_owner TO
                                        SHARE share_poc;
```

This results in this error: "Future grant on objects of type TABLE to SHARE is restricted."

Instead, you must GRANT entitlement *after* object creation as shown later within this section.

Grant entitlement to a new role to use existing warehouses:

```
GRANT USAGE ON WAREHOUSE tpc_wh_xsmall TO ROLE share_owner_role;
GRANT USAGE ON WAREHOUSE tpc_wh_small  TO ROLE share_owner_role;
GRANT USAGE ON WAREHOUSE tpc_wh_medium TO ROLE share_owner_role;
GRANT USAGE ON WAREHOUSE tpc_wh_large  TO ROLE share_owner_role;
GRANT USAGE ON WAREHOUSE tpc_wh_xlarge TO ROLE share_owner_role;
```

Assign the share to the desired consumer account, noting there may be several consuming accounts requiring service.

We will use your managed account to import your share.

```
USE ROLE accountadmin;
```

Replace <SHARE ACCOUNT> in the next SQL statement with the JSON accountLocator or locator attribute from your managed account created earlier.

```
ALTER SHARE share_poc ADD ACCOUNTS = <SHARE ACCOUNT>;
```

Noting your locator will differ, mine is as follows:

```
ALTER SHARE share_poc ADD ACCOUNTS = HR83528;
```

331

All SQL statements within this section must be rerun for each new consumer where bespoke object content is created.

Unentitled Objects

As the heading suggests, these objects are not entitled and can be passed straight through to your share, bypassing the entitlement engine. Figure 9-8 illustrates how you create and then share a new object called v_region.

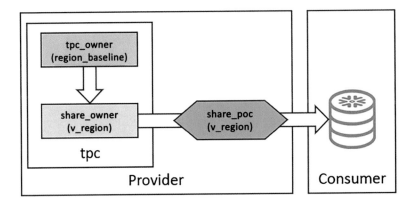

Figure 9-8.   *Unentitled object share*

First, you set your execution context.

```
SET share_owner_role   = 'share_owner_role';
SET tpc_database       = 'tpc';
SET share_owner_schema = 'tpc.share_owner';
SET tpc_warehouse_XS   = 'tpc_wh_xsmall';

USE ROLE      IDENTIFIER ( $share_owner_role   );
USE DATABASE  IDENTIFIER ( $tpc_database        );
USE SCHEMA    IDENTIFIER ( $share_owner_schema );
USE WAREHOUSE IDENTIFIER ( $tpc_warehouse_xs   );
```

Reference data is a common use case for passthrough objects; in this example you will create a secure view.

```
CREATE OR REPLACE SECURE VIEW v_region COPY GRANTS
AS
SELECT  r_regionkey,
        r_name,
        r_comment
FROM    tpc.tpc_owner.region_baseline;
```

The share_owner database may contain objects sourced from many different databases and schemas. The share_owner database is intended to be the container from which your share is populated; you should containerize your objects for ease of later maintenance and administration.

You cannot directly entitle your share called share_poc to access the new secure view v_region_baseline using the role share_owner_role. Attempting to do so results in the following error:

> ### Share '"<YOUR ACCOUNT>.SHARE_POC'" does not exist or not authorized.

You must first switch the role.

```
USE ROLE securityadmin;
```

Then use GRANT SELECT on individual views.

```
GRANT SELECT ON tpc.share_owner.v_region TO SHARE share_poc;
```

An alternative approach is to create all the desired nonview objects and then entitle all of them using a single command per object type.

```
GRANT SELECT ON ALL TABLES             IN SCHEMA tpc.share_owner TO SHARE
                                       share_poc;
GRANT SELECT ON ALL DYNAMIC     TABLES IN SCHEMA tpc.share_owner TO SHARE
                                       share_poc;
GRANT SELECT ON ALL MATERIALIZED VIEWS IN SCHEMA tpc.share_owner TO SHARE
                                       share_poc;
```

To confirm entitlement and objects granted to your share share_poc, use this:

```
SHOW GRANTS TO SHARE share_poc;
```

Importing a Share

Once your share has been populated and entitled for a consuming account, you must log in to your new managed account. In this example, the URL is `https://hr83528.eu-west-2.aws.snowflakecomputing.com`; yours will differ.

Using the managed account credentials repeated next, log in to your managed account.

```
ADMIN_NAME     = 'poc_admin'
ADMIN_PASSWORD = 'POC_admin_123'
```

Importing shares is performed by the `ACCOUNTADMIN` role:

```
USE ROLE accountadmin;
```

```
SHOW SHARES IN ACCOUNT;
```

You should see two inbound shares. Note that `SNOWFLAKE` is provided by Snowflake Inc. You should also see `database_name` is unpopulated for the new share. You create a database (see Figure 9-9).

created_on	kind	owner_account	name	database_name
2024-01-22 02:13:55.631 -0800	INBOUND	ACXELCQ.ZI95050	SHARE_POC	
2021-01-25 17:57:04.733 -0800	INBOUND	SNOWFLAKE	ACCOUNT_USAGE	SNOWFLAKE

Figure 9-9. *Inbound share listing*

You now create a database from the inbound share. Note that the owner account and share name will differ from yours.

```
CREATE DATABASE share_poc_database
FROM   SHARE   ACXELCQ.ZI95050.share_poc;
```

Now check that the `database_name` attribute is populated for your share.

```
SHOW SHARES IN ACCOUNT;
```

You should also see the new database listed in the database browser when refreshed, as shown in Figure 9-10.

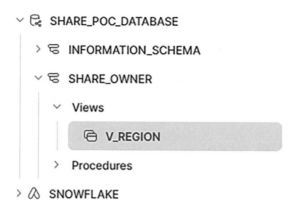

Figure 9-10. *Imported database listing*

Now create a warehouse.

```
CREATE OR REPLACE WAREHOUSE poc_wh WITH
WAREHOUSE_SIZE      = 'X-SMALL'
AUTO_SUSPEND        = 60
AUTO_RESUME         = TRUE
MIN_CLUSTER_COUNT   = 1
MAX_CLUSTER_COUNT   = 4
SCALING_POLICY      = 'STANDARD'
INITIALLY_SUSPENDED = TRUE;
```

Set your session context to use your newly created warehouse.

```
USE WAREHOUSE poc_wh;
```

And later test your imported database once v_region has been provisioned (next) to ensure you can SELECT data.

```
SELECT r_name,
       r_comment
FROM   share_poc_database.share_owner.v_region;
```

Imported databases are owned by the ACCOUNTADMIN role by default. Your client must create their own roles and grant entitlements for their internal use.

Entitled Data Sharing

You will extend the newly created share `share_poc` to include entitled objects that cannot pass straight through to your share. The entitlement engine curates the data content of a shared object acting as the filter engine.

Designing a Filter Engine

As you have seen, sharing unentitled objects is relatively simple. Sharing entitled objects, that is, a subset of data contained within an object, is not simple.

Creating a limited scope number of bespoke objects for an individual client is easy. When the limited scope changes, you will find yourself overwhelmed with demand. Therefore, adopting a pattern-based approach to generating bespoke objects is the only viable way forward.

The first decision is to determine where to build your entitlement engine, and for the purposes of this example, you will reuse the `tpc_owner` schema. For a real-world implementation, you may choose to develop your entitlement engine within a separate schema.

Filter Engine Requirements

For simplicity's sake you will reuse the table `base_customer_order` from a previous chapter as the data source for generating customer-specific filtered data.

Table 9-1 shows the table `base_customer_order` definition.

Table 9-1. *base_customer_order Table Definition*

Attribute Name	Datatype
customer_key	NUMBER (38,0)
customer_name	VARCHAR(25)
order_key	NUMBER (38,0)
order_date	DATE
nation_name	VARCHAR(25)
region_name	VARCHAR(25)

Let's assume you are required to do the following:

- Generate an object containing a client-specific view of data

- Filter by `region_name` to generate only "Africa" data

- Mask the `order_key` to prevent identification of individual orders

Over time you can expect to have multiple clients with both differing region filters and differing masking requirements. I assume your source data product tables and views contain a complete superset of data.

Filter Engine Model

Figure 9-11 shows the client entitlement, filter engine, and data product model.

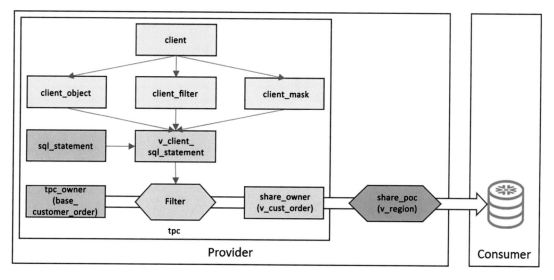

Figure 9-11. *Filter engine design*

You will now define each entity and focus on the minimal attributes to develop a simple data model. Feel free to extend them according to your needs later.

While Snowflake does not enforce referential integrity, I will show how to implement referential integrity to demonstrate the capability, because the optimizer may use referential integrity to inform internal decision-making. By convention you use a sequence per object to generate a surrogate primary key.

Filters and masks use substitution variables enclosed in square brackets, [ ]. These are evident within the corresponding data.

Client

Clients are the consumers of curated data products; in real-world use, you may find clients are mastered elsewhere and then fed along with entitlements into the data model. Clients are the entry point to your model and later in this section will be used to drive the generation of bespoke content.

First set your execution context.

```
SET tpc_owner_role    = 'tpc_owner_role';
SET tpc_database      = 'tpc';
SET tpc_owner_schema  = 'tpc.tpc_owner';
SET tpc_warehouse_XS  = 'tpc_wh_xsmall';

/* Set execution context */
USE ROLE      IDENTIFIER ( $tpc_owner_role   );
USE DATABASE  IDENTIFIER ( $tpc_database     );
USE SCHEMA    IDENTIFIER ( $tpc_owner_schema );
USE WAREHOUSE IDENTIFIER ( $tpc_warehouse_xs );
```

Then create a table called `client` along with sequence `seq_client_id` to generate surrogate keys.

```
CREATE OR REPLACE TABLE client
(
client_id       NUMBER    PRIMARY KEY,
name            VARCHAR(255),
account         VARCHAR(255)
);

CREATE OR REPLACE SEQUENCE seq_client_id START WITH 100000;
```

Now create your first client and extend the clients to suit your needs.

```
INSERT INTO client VALUES
( seq_client_id.NEXTVAL, 'POC', 'Proof of Concept Client' );
```

Client Object

Objects refer to the physical objects holding data or functionality, which your clients have either purchased or licensed. Objects hold the superset of data products that you seek to monetize.

Create a table called client_object along with sequence seq_client_object_id to generate surrogate keys.

```
CREATE OR REPLACE TABLE client_object
(
client_object_id    NUMBER,
client_id           NUMBER   REFERENCES client ( client_id ),
object_name         VARCHAR(255)
);

CREATE OR REPLACE SEQUENCE seq_client_object_id START WITH 100000;
```

Assign a single database object base_customer_order to your client.

```
INSERT INTO client_object
SELECT seq_client_object_id.NEXTVAL,
       ( SELECT client_id FROM client WHERE name = 'POC' ),
       'base_customer_order';
```

Client Filter

Filters refer to the object physical attributes holding data you want to filter on. Equivalent to row-level security (RLS), this section enables data subsetting at object generation time. You prefer to not use RLS when generating client-specific objects in order to do the following:

- Reduce replicated data to a minimum

- Prevent resolving RLS for every SQL call made to the client object

- Remove the need to create RLS policies "on the fly"

Create a table called client_filter along with sequence seq_client_filter_id to generate surrogate keys.

```
CREATE OR REPLACE TABLE client_filter
(
client_filter_id    NUMBER,
client_id           NUMBER   REFERENCES client ( client_id ),
filter_name         VARCHAR(255),
filter_attribute    VARCHAR(255),
filter_value        VARCHAR(255)
);

CREATE OR REPLACE SEQUENCE seq_client_filter_id START WITH 100000;
```

Apply a single filter to base_customer_order.region_name for AFRICA.

```
INSERT INTO client_filter
SELECT seq_client_filter_id.NEXTVAL,
       ( SELECT client_id FROM client WHERE name = 'POC' ),
       '[REGION]',
       'region_name',
       'AFRICA';
```

Client Mask

Masks refer to the object physical attributes holding data you want to mask. Equivalent to data masking, this section enables data masking at object generation time.

Create a table called client_mask along with a sequence called seq_client_mask_id to generate surrogate keys.

```
CREATE OR REPLACE TABLE client_mask
(
client_mask_id      NUMBER,
client_id           NUMBER   REFERENCES client ( client_id ),
mask_name           VARCHAR(255),
mask_attribute      VARCHAR(255),
mask_value          VARCHAR(255)
);

CREATE OR REPLACE SEQUENCE seq_client_mask_id START WITH 100000;
```

Apply a single filter to base_customer_order.order_key, setting the value to
********.

```
INSERT INTO client_mask
SELECT seq_client_mask_id.NEXTVAL,
       ( SELECT client_id FROM client WHERE name = 'POC' ),
       '[ORDER_KEY_MASK]',
       'order_key',
       '''********''';
```

Denormalize Client Information

With all your client entities both created and populated, you should denormalize the
client-specific components to make your data model more easily understood and user
friendly. You do this by creating a view called v_client_info to join all client-specific
tables together.

```
CREATE OR REPLACE VIEW v_client_info
AS
SELECT c.name     AS client_name,
       c.account  AS client_account,
       co.object_name,
       cf.filter_name,
       cf.filter_attribute,
       cf.filter_value,
       cm.mask_name,
       cm.mask_attribute,
       cm.mask_value
FROM   client        c,
       client_object co,
       client_filter cf,
       client_mask   cm
WHERE  c.client_id   = co.client_id
AND    c.client_id   = cf.client_id
AND    c.client_id   = cm.client_id;
```

Next check you have a single record.

```
SELECT * FROM v_client_info;
```

SQL Statement

SQL statements overlay data product objects providing the template for client-specific data generation into client objects. For this example, you set the sql_statement name to be the same as the client_object name in order to later join the data.

Create a table called sql_statement along with a sequence called seq_sql_statement_id to generate surrogate keys.

```
CREATE OR REPLACE TABLE sql_statement
(
sql_statement_id   NUMBER    PRIMARY KEY,
name               VARCHAR(255),
sql_statement      VARCHAR(255)
);

CREATE OR REPLACE SEQUENCE seq_sql_statement_id START WITH 100000;
```

Define a SQL statement with substitution values for a filter and mask, noting you set the name to be base_customer_order to join with the client configuration data.

```
INSERT INTO sql_statement VALUES
( seq_sql_statement_id.NEXTVAL, 'base_customer_order',
  'SELECT customer_key, customer_name, [ORDER_KEY_MASK], order_date, nation_
  name FROM base_customer_order WHERE region_name = ''[REGION]'' ' );
```

Client SQL Statement View

With all the entities defined, you now bring them together into a single usable entitlement generation object.

Create a view called v_client_sql_statement.

```
CREATE OR REPLACE VIEW v_client_sql_statement
AS
SELECT c.client_name,
```

```
      REPLACE ( REPLACE ( s.sql_statement, '[REGION]', c.filter_value ),
      '[ORDER_KEY_MASK]', c.mask_value ) AS client_sql_statement,
      c.client_account,
      c.object_name,
      c.filter_name,
      c.filter_attribute,
      c.filter_value,
      c.mask_name,
      c.mask_attribute,
      c.mask_value,
      s.sql_statement
FROM  v_client_info    c,
      sql_statement    s
WHERE c.object_name    = s.name;
```

Now check that the view v_client_sql_statement returns the expected results.

```
SELECT * FROM v_client_sql_statement;
```

You should see attribute client_sql_statement returns the next SQL statement, noting I have appended LIMIT 10 to restrict the returned result set.

```
SELECT customer_key, customer_name, '********',
      order_date, nation_name
FROM  base_customer_order
WHERE region_name = 'AFRICA'
LIMIT 10;
```

With your view v_client_sql_statement prepared, you are ready to build your filter engine.

Building a Filter Engine

Building a filter engine brings together several components:

- Creation of containers to hold client-specific curated objects

- Creation of objects within a schema

- Granting entitlement on schema objects to the share

As with all code, full testing should be conducted and signed off on before scheduled deployment.

We do not advocate the automated deployment of generated code.

You can now build a JavaScript stored procedure to generate your code passing through a single parameter called P_CLIENT_NAME to generate the client-specific containers and objects.

```
CREATE OR REPLACE PROCEDURE sp_create_share ( P_CLIENT_NAME STRING )
RETURNS string
LANGUAGE javascript
EXECUTE AS CALLER
AS
$$
   var sql_stmt  = "";
   var recset    = "";
   var err_state = "";
   var result    = "";

   var client_account = "";
   var share_grants   = "";

   result  = "/* Create a role to manage shared objects */\n"
   result += "USE ROLE accountadmin;\n"
   result += "CREATE SHARE IF NOT EXISTS share_" + P_CLIENT_NAME + ";\n\n"

   result += "/* Create a schema for shared objects */\n"
   result += "USE ROLE sysadmin;\n"
   result += "CREATE SCHEMA IF NOT EXISTS tpc.share_owner_" + P_CLIENT_NAME
   + ";\n\n"

   result += "/* Entitle new share to access new schema */\n"
   result += "GRANT USAGE            ON DATABASE tpc
   TO SHARE share_" + P_CLIENT_NAME + ";\n"
   result += "GRANT REFERENCE_USAGE ON DATABASE tpc
   TO SHARE share_" + P_CLIENT_NAME + ";\n"
```

```
result += "GRANT USAGE
ON SCHEMA    tpc.share_owner_" + P_CLIENT_NAME + " TO SHARE share_" +
P_CLIENT_NAME + ";\n\n"

result += "/* Entitle new role to create objects in the new schema */\n"
result += "GRANT USAGE
ON SCHEMA tpc.share_owner_" + P_CLIENT_NAME + " TO ROLE share_owner_
role;\n"
result += "GRANT CREATE TABLE
ON SCHEMA tpc.share_owner_" + P_CLIENT_NAME + " TO ROLE share_owner_
role;\n"
result += "GRANT CREATE VIEW
ON SCHEMA tpc.share_owner_" + P_CLIENT_NAME + " TO ROLE share_owner_
role;\n"
result += "GRANT CREATE MATERIALIZED VIEW
ON SCHEMA tpc.share_owner_" + P_CLIENT_NAME + " TO ROLE share_owner_
role;\n"
result += "GRANT CREATE DYNAMIC TABLE
ON SCHEMA tpc.share_owner_" + P_CLIENT_NAME + " TO ROLE share_owner_
role;\n\n"

/* Fetch client curated objects */
sql_stmt  = "SELECT client_account,\n"
sql_stmt += "        object_name,\n"
sql_stmt += "        client_sql_statement\n"
sql_stmt += "FROM    v_client_sql_statement\n"
sql_stmt += "WHERE   client_name = :1;"

stmt = snowflake.createStatement( { sqlText: sql_stmt, binds:[P_CLIENT_
NAME] } );

try
{
   recset = stmt.execute();
   while(recset.next())
   {
      client_account = recset.getColumnValue(1);
```

```
        result += "CREATE OR REPLACE VIEW tpc.share_owner_" + P_CLIENT_
        NAME + "." + recset.getColumnValue(2) + "\n"
        result += "AS\n"
        result += recset.getColumnValue(3) + ";\n\n"

        share_grants += "GRANT SELECT ON tpc.share_owner_" + P_CLIENT_NAME
        + "." + recset.getColumnValue(2) + " TO SHARE share_" + P_CLIENT_
        NAME + ";\n"
    }
}
catch(err)
{
    err_state += "\nFail Code: " + err.code;
    err_state += "\nState: " + err.state;
    err_state += "\nMessage: " + err.message;
    err_state += "\nStack Trace:" + err.StackTraceTxt;
    err_state += "\nSQL Statement:\n\n" + result;
    result = err_state;
}

result += "/* Entitle new objects to share */\n"
result += "USE ROLE securityadmin;\n"
result += share_grants + "\n";

result += "/* Entitle new role to access objects in both existing schema
and new schema */\n"
result += "GRANT SELECT ON ALL TABLES
IN SCHEMA tpc.share_owner_" + P_CLIENT_NAME + " TO ROLE share_owner_
role;\n"
result += "GRANT SELECT ON ALL DYNAMIC TABLES
IN SCHEMA tpc.share_owner_" + P_CLIENT_NAME + " TO ROLE share_owner_
role;\n"
result += "GRANT SELECT ON ALL MATERIALIZED VIEWS
IN SCHEMA tpc.share_owner_" + P_CLIENT_NAME + " TO ROLE share_owner_
role;\n\n"

result += "/* Make share available to consumer account */\n"
```

```
result += "USE ROLE accountadmin;\n"
result += "ALTER SHARE share_" + P_CLIENT_NAME + " ADD ACCOUNTS = '" +
client_account + "';\n\n"

return result;
$$;
```

Call sp_create_share with your client POC.

```
CALL sp_create_share ( 'POC' );
```

The sp_create_share should return the following SQL statements noting the inline comments to explain each section.

```
/* Create a role to manage shared objects */
USE ROLE accountadmin;
CREATE SHARE IF NOT EXISTS share_POC;

/* Create a schema for shared objects */
USE ROLE sysadmin;
CREATE SCHEMA IF NOT EXISTS tpc.share_owner_POC;

/* Entitle new share to access new schema */
GRANT USAGE             ON DATABASE tpc                 TO SHARE share_POC;
GRANT REFERENCE_USAGE ON DATABASE tpc                 TO SHARE share_POC;
GRANT USAGE             ON SCHEMA   tpc.share_owner_POC TO SHARE share_POC;

/* Entitle new role to create objects in the new schema */
GRANT USAGE                     ON SCHEMA tpc.share_owner_POC TO ROLE share_
                                owner_role;
GRANT CREATE TABLE              ON SCHEMA tpc.share_owner_POC TO ROLE share_
                                owner_role;
GRANT CREATE VIEW               ON SCHEMA tpc.share_owner_POC TO ROLE share_
                                owner_role;
GRANT CREATE MATERIALIZED VIEW ON SCHEMA tpc.share_owner_POC TO ROLE share_
                                owner_role;
GRANT CREATE DYNAMIC TABLE      ON SCHEMA tpc.share_owner_POC TO ROLE share_
owner_role;

CREATE OR REPLACE SECURE VIEW tpc.share_owner_POC.base_customer_order
```

```
AS
SELECT customer_key, customer_name, '********', order_date, nation_name
FROM base_customer_order WHERE region_name = 'AFRICA';

/* Entitle new objects to share */
USE ROLE securityadmin;
GRANT SELECT ON tpc.share_owner_POC.base_customer_order TO SHARE share_POC;

/* Entitle new role to access objects in both existing schema and new
schema */
GRANT SELECT ON ALL TABLES            IN SCHEMA tpc.share_owner_POC TO
                                      ROLE share_owner_role;

GRANT SELECT ON ALL DYNAMIC TABLES    IN SCHEMA tpc.share_owner_POC TO
                                      ROLE share_owner_role;

GRANT SELECT ON ALL MATERIALIZED VIEWS IN SCHEMA tpc.share_owner_POC TO
                                      ROLE share_owner_role;

/* Make share available to consumer account */
USE ROLE accountadmin;
ALTER SHARE share_POC ADD ACCOUNTS = 'ABC123';
```

Deploying Generated Code

The following stored procedure sp_create_share can be extended in several ways. For example:

- Writing output to a logging table

- Building tables, not views

- Generating test cases

Regardless of the actual code generated, I strongly recommend full testing is conducted with business sign-off before deployment.

Setting the Standard

Having discussed entitling your data products, let's discuss how you can set the standard for distributing your data products. I will not discuss this in great depth but will call out some available options.

Imported Database Entitlement

Imported databases created from shares or replicated databases do not import source provider entitlement. By now you should all be familiar with role-based access control (RBAC), you should provision a sample RBAC script segregating your shared objects into data products for your client to begin integrating your data products with their local data sets.

The sample RBAC should be accompanied by a data model and data catalog.

Sample SQL for Common Use Cases

Regardless of whether your application has implemented embedded entitlement or pre-filtered entitlement, you should provision a suite of SQL statements. These SQL statements should implement common use cases and act as a starting point for your clients by demonstrating how to extract business value from the data product.

A suite of tuned, performant example SQL statements will help remediate client performance issues by demonstrating both functional and performant Snowflake interaction. You must remember that the Snowflake optimizer functions differently than legacy relational database management systems (RDBMSs). Therefore, your sample SQL statements will also serve as a guide to uplift your client skills.

Client Collaboration

Many clients use your organization's data products in conjunction with both your competitor data products and their own internal data products. Joining across imported schemas where embedded entitlement logic exists in third-party data sets is highly likely to result in performance issues.

I advise caution. Before purchasing a third-party data set, I suggest you fullly understand all of the underlying query object implementations.

Clients are rightly protective of their intellectual property and in the event of a performance issue, the data product provider will prefer access to the exact queries issued by the consumer. However, sharing SQL is problematic; SQL often contains bespoke logic, and consumers must be prevented from accessing a provider's intellectual property.

Historized Data

Many data products are offered as point-in-time, current view only. That is, ingested data overwrites earlier data without retaining a history. Ingested data may not contain every intermediate transaction recorded by a source system; therefore, each uploaded file contains a snapshot.

An easy-to-implement value-add is to retain the full history of all captured data. By adding temporal attributes to record valid-from and valid-to date stamps, you can provide the ability to reconstruct data at any point in time for the retained data.

You can find more information on slowly changing dimensions at `https://en.wikipedia.org/wiki/Slowly_changing_dimension`.

Implementing temporal attributes offers an inexpensive approach to adding value to your data particularly when generating your codebase.

Data Model

At some point your clients may want your data product to be modeled in a particular form so that it's compatible with their existing implementation. The first step toward integrating your data product with client data model is to publish an entity-relationship diagram.

From a data provider's perspective, offering consistent and interlinked data models across all data product offerings is a worthy ambition. But this is hard to achieve in practice. I do not prescribe any particular data modeling technique except to say 3NF, DV2.0, and the dimensional approach each offers advantages and disadvantages.

Adopting an `INSERT`-only data model provides the fastest method of delivering data into shared objects.

Data Catalog

Data catalogs articulate both the business meaning and the technical metadata for each entity and attribute within the data model. Providing contextual information is always a good thing to do, and your clients will value knowing the provider supplied meaning to enrich their own understanding and flesh out their own data catalog.

Large organizations find it difficult to articulate their whole data product shop window using a single tool, and many organizations have incomplete catalogs. Provide what you have while completing the remainder. Your clients both expect and appreciate all available information.

Shared Tag References

The ability to export data product context is becoming more important to clients. Data without context is akin to *data littering*, a term coined by Dr Jon Talburt and referenced within this article: `https://tdan.com/data-speaks-for-itself-data-littering/29122`.

At the time of writing, shared tag references are a public preview feature for which further information can be found at `https://docs.snowflake.com/en/user-guide/data-sharing-provider#shared-tag-references`.

Multiple Shares of Same Data

A common requirement from clients is to have the same data shared into multiple accounts. Where the consumer account is co-located with the provider account, the data does not move. For clients sharing objects access the same micro-partitions as the provider, there is zero cost.

From a consumer perspective, with a single share, accessing the same data set in multiple environments can be achieved only by replicating the reshared data, as shown in Figure 9-12.

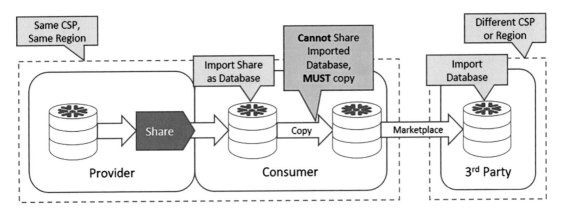

Figure 9-12. *Sharing limitation*

Data shared within the same CSP and same region is readily achieved by enabling the consuming account to access a single share. Offering multiple account consumption for an existing share is a great way to win customer loyalty and significantly reduce both friction and copy costs.

Hydration Approach

Regardless of where you distribute your data products, you must ensure the consistent application of entitlement across disparate data distribution venues. Your approach must consider not just Snowflake, but all tooling and products dependent upon data products mastered within Snowflake.

To protect your clients from internal system failure, you must also consider how, and from where, you will hydrate your data. This point may not be obvious: your clients should not be forced to invoke their disaster recovery plan due to an upstream data supply failure from a provider. Furthermore, you should consider hydrating from multiple sources wherever possible.

The timeliness of data must also be considered. If you curate your data within a single environment and propagate change to consumers, with the exception of a Secure Direct data share, latency is introduced. How you address latency is important; in the previous chapter you saw how parallelizing feeds can reduce latency.

Summary

This chapter by considering how clients interact with your data products, and you saw some of the consumption challenges your clients may face. I explained two different approaches to entitling data along with the pros and cons of each.

After explaining how to pass through unentitled objects, we decided upon delivering curated data sets for each client. You then saw how to develop a simple code generator delivering bespoke content according to your configured entitlement model.

The example illustrated within this chapter was intended to introduce how client-specific entitled data can be both derived and delivered in a semi-automated manner. The code was simplistic in its approach and will not suffice for more complex requirements, and I encourage experimentation; you may want to add both entities and attributes to extend filter engine capability. Furthermore, a combination of access policies and data masking may suffice instead of data-driven curated data sets. Your use cases will inform your decision-making.

The final section addressed some common challenges for which you can either resolve very easily or initiate a conversation with your data product providers to identify better solutions.

Views are not the only sharable object type. The Snowflake documentation at `https://docs.snowflake.com/en/user-guide/data-sharing-intro` shows that the following object types are sharable:

- Tables

- External tables

- Dynamic tables

- Secure views

- Secure materialized views

- Secure UDFs

The list of sharable object types has changed since writing my previous book, *Maturing the Snowflake Data Cloud*, with the addition of dynamic tables.

You can now move on to the final chapter in this book, which covers what to look out for and how to approach optimizing performance.

CHAPTER 10

Optimizing Performance

In previous chapters, you investigated Snowflake performance tuning from different perspectives and established a clear understanding through practical investigation and hands-on examples. This final chapter brings all of your learnings together into one place with the intention of providing a pragmatic guide to aid in your future investigations.

I do not claim to cover 100 percent of all possible scenarios or Snowflake performance issues in this chapter. With ever-expanding platform capability, Snowflake continually finds ways to improve performance, improve available information, and deliver tooling to improve processes. Its goals are to provide a starting point for investigation and provide contextual information to open up new pathways for investigation.

I have attempted to make this chapter light on code but provide template code samples and summary queries. Many expanded code examples are contained in previous chapters; my hope is that this book will be well used over time!

Tuning must be regarded as a continual activity. Treat the root cause, not the symptoms!

Naturally some information overlaps occur when investigating performance issues; this chapter is no exception. There is no single "right path" to begin an investigation. Your entry point will depend upon what you already know and the context, whether planning a new application, investigating a newly reported issue, or remediating a known issue. Intersections between different investigation paths offer insight into new avenues, possibilities, and opportunities for both learning and improvements to be made. I encourage both investigation and experimentation. You will often learn more through failure than success!

© Andrew Carruthers 2024
A. Carruthers, *Tuning the Snowflake Data Cloud*, https://doi.org/10.1007/979-8-8688-0379-6_10

Snowsight provides a window into Snowflake performance but does not provide the deep drill-down capability exposed in this book. I hope your efforts empower you to both up-skill and deliver impactful business success by reducing costs and improving your application code performance.

Let's start by looking at design decisions made before a line of code is crafted.

Early Design Decisions

Design decisions made during the early stages of platform choice often have a decisive impact on system performance. In this section, I discuss some points to consider when implementing Snowflake.

As noted in Chapter 1, tuning the design is the most effective way of achieving optimal performance. This step occurs *before* attempting to write any code. The same advice applies when working with any technology, not just Snowflake. There are many ways to implement poor designs and far fewer ways to implement good designs.

As Tony Robbins observes, "Complexity is the enemy of execution." We should strive to reduce complexity at every opportunity.

Snowflake Edition Costs

The edition of Snowflake you choose can have material impact on both cost and feature availability. Throughout this book and for previous books in this series, I have recommended all trial accounts be created using Business Critical Edition to ensure the most complete feature set is available for use as you investigate Snowflake. But Business Critical Edition is not the only option; compute costs per credit correlate to the Snowflake edition chosen. You can find more information at `https://www.snowflake.com/en/data-cloud/pricing-options/`.

When considering Snowflake, you must identify the minimum platform feature set required to support your intended use case. Your applications may not use or require every Snowflake feature, and you are wise to consult your cybersecurity colleagues for their input before deciding on a Snowflake edition. I consider Business Critical Edition to be the optimal edition for its security profile and extended capability for continuous data protection (failover and client redirect).

Please refer to the feature list tables and overview of the Snowflake core platform capabilities by edition found here: `https://docs.snowflake.com/en/user-guide/intro-editions?utm_source=snowscope&utm_medium=serp&utm_term=edition#feature-edition-matrix`.

Data Model Approach

Snowflake is "model agnostic." Third normal form, Data Vault 2.0, and star schema all work very well. You must understand your data volume, velocity, and variety in order to decide upon a data modeling approach. Choosing the wrong model for your data profile can adversely affect performance and increase costs. In general, you know from earlier investigations conducted in this book that both `UPDATE` and `MERGE` operations are expensive, whereas `INSERT` and many (but not all) `DELETE` operations are relatively inexpensive. You must balance your approach with the need to query data. Because reading data is performed far more often than writing data, tuning `SELECT` statements is always worthwhile.

Where your application requires historized or bi-temporal data, adopting an insert-only pattern such as Data Vault 2.0 will provide optimal performance throughout the application life cycle.

Snowflake performs best with high-volume, low-frequency data operations and is less performant with low-volume, high-frequency data operations. Transactional workloads should be avoided when not using hybrid tables. Note that any use of hybrid tables should be limited to low millions of records according to conversation with the Snowflake Sales Support engineering staff. Mitigation by partitioning workloads as described earlier in this book can be very effective.

Regardless of the data modeling approach, avoid forcing object and attribute names to either mixed or lowercase. I prefer most objects and attributes names to remain in uppercase, preventing objects and attributes from being referenced in double quotes in SQL statements. Procedure and function names may benefit from mixed-case naming according to preference.

Platform Differences

A full treatise on migrating from disparate legacy platforms to Snowflake is beyond the scope of this book. Various guides exist to begin the migration process, though not all the details are covered. You can find a good starting point at `https://community.snowflake.com/s/article/So-You-Want-to-Migrate-to-Snowflake-Part-One`.

Some additional considerations also apply; note that the following list is incomplete:

- Oracle incorrectly implements `NULL` logic according to the ANSI standard, whereas Snowflake implements `NULL` correctly.

- Never use `SELECT *` in Snowflake; always declare every attribute even if all attributes from the table are used in the wider query.

- Snowflake does not support physical table partitioning, though a similar effect can be achieved with parallelizing operations, as described in Chapter 8.

- Not all legacy RDBMS implement ACID transactions by default. For some, the isolation level must be set to block writes to prevent dirty reads.

In Chapter 1 I discussed migration guides and listed several common legacy RDBMS platform guides. I also noted the availability of SnowConvert and detailed alternative options. You can find more information on migration kits at `https://www.snowflake.com/migrate-to-the-cloud/`.

Logging

All applications require process metadata to be stored to trace the inevitable feed ingestion issues arising during the course of day-to-day operation.

A single logging table effectively serializes all concurrent parallel processing due to the immutable nature of micro-partitions and locking operation where constant micro-partition churn occurs. My recommendation is to use an Event table as described previously, which removes the serialization issue. Note that a latency of a few minutes is commonly experienced between event creation and event observability.

The forthcoming Unistore workload uniting transactional and analytical data using hybrid tables may provide a single logging table capability without blocking parallel transactions. We have not proven Unistore workloads in this book, but I bring this to your attention for future reference. You can find more details on Unistore at `https://www.snowflake.com/en/data-cloud/workloads/unistore/`.

Role-Based Access Control

When implemented optimally, Snowflake role-based access control (RBAC) provides an excellent approach to securing objects, attributes, and data. When poorly implemented by nesting several layers of roles, performance issues arise due to the optimizer drilling down through each layer and through view definitions to determine object access entitlement.

Snowflake provides core administrative roles that may be wrapped for both single-tenant and multitenant environments, a practice discussed in *Maturing the Snowflake Data Cloud*. In addition to the Snowflake-supplied roles, you should define separate roles for the following:

- Application-owned objects

- Data manipulation in application objects

Segregating object ownership from object usage is critical. Each logical grouping of objects should have its own ownership role. Furthermore, you should implement data manipulation (or object usage) roles according to their function, as shown in Figure 10-1.

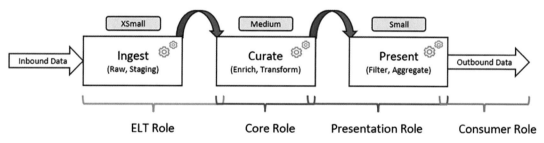

Figure 10-1. *Data manipulation roles*

These are the application roles:

- **ELT role:** Used to ingest data into our application; scope is limited to staging/raw table population and triggering functionality to begin the curation process.

- **Core role:** Used to perform all curation activity to build our data products in preparation for presentation.

- **Presentation role:** Used to implement client-specific data sets and data sharing capability; references curated data products.

- **Consumer role:** There may be many consumer roles according to end-user requirements. These roles are for directly connected users and reference specific presentation objects.

A multitenant environment will have bespoke application roles for each tenant. You should be mindful of tooling, which purports to simplify RBAC. I strongly caution against those tools that implement a role per user and that insist all RBAC is defined in their tooling for management. These issues can arise:

- Role proliferation will cause performance issues when resolving object entitlement.

- Over-enthusiastic application of data masking policies will result in slow metadata operations.

- Vendor lock-in will occur where a product exclusively encapsulates RBAC management.

We recommend roles are kept to a minimum with few nested layers and that data masking policies are applied sparingly.

Just because a feature is available does not mean the feature should be used ubiquitously.

Declare Constraints

Snowflake allows referential integrity to be declared but not enforced. With the exclusion of hybrid tables, the only constraint enforced is NOT NULL. You can find more information at https://docs.snowflake.com/en/sql-reference/constraints-overview.

The presence of unenforced constraints may influence query optimizer processing and greatly assist self-discovery by third-party tooling. I therefore recommend that constraints are declared wherever possible.

Transient or Permanent Tables?

I am assuming that inbound data feeds are repeatable and therefore suggest ingestion of raw or staging tables should use transient tables with Time Travel set to 0 as transient tables do not utilize Fail-Safe. You might also consider using transient tables for frequently refreshed data generated for point-in-time reporting.

Permanent tables should be reserved for persistent storage where Time Travel is required. Note that the seven-day Fail-Safe period follows. Typically, permanent tables are used for storing both curated and historized data.

Parallelizing data loads as shown in Chapter 8 can significantly reduce micro-partition churn for permanent tables.

Warehouse Considerations

Correctly sizing and scaling a warehouse is dependent upon a full understanding of workload under "steady state" conditions. Previously in Chapter 8 we asked these questions:

- Is queueing, spills to disk, or OOMs evident?
- Is the warehouse overloaded, queueing, or blocking?
- Is the data feed overrunning its schedule leading to feeds backing up?
- Are costs increasing over time?

Later in this chapter I summarize queries to answer each question noting that there are several ways to answer these questions.

As discussed in Chapter 6, I prefer fixed-size warehouses where the warehouse declaration remains constant.

I do not advocate dynamically resizing warehouses. This is a poor approach to performance tuning.

A known, fixed suite of declared warehouses is preferable to dynamically managed warehouses. Dynamically resizing warehouses disables the warehouse cache.

Additionally, I advocate consolidating warehouses of the same size into a single declaration. I also propose query tags with JSON to differentiate usage as discussed next. You also want minimal warehouse lag. Suspending and resuming warehouses introduce latency, and some use cases benefit from keeping warehouses "warm." Conversely, starting and stopping warehouses in an ad hoc manner can result in under-utilization and additional spend, which delivers no value.

Workload Monitoring

Where workloads are consolidated into generic warehouses, consumption metrics are more difficult to attribute to the consuming source. To maintain the efficiencies gained by consolidating workloads, query tags should be set before a SQL statement is issued, and unset, or set to a new value for subsequent SQL statements.

You should set individual query tags for every SQL operation in your system. The use of query tags provides a very fine grain of traceability back to the source when investigating performance issues.

I strongly recommend implementing query tags to assist in later investigations.

An individual query tag may contain up to 2,000 characters and can contain JSON.

```
ALTER SESSION SET query_tag =  '{"Team": "Finance", "Query":
"BusinessLineYTD"}';
```

You can investigate query tag values using the SHOW command.

```
SHOW PARAMETERS LIKE 'query_tag';
```

Then extract the "value" programmatically.

```
SELECT "key",
       "value"
FROM   TABLE ( RESULT_SCAN ( last_query_id()));
```

Likewise, you can unset a query tag.

```
ALTER SESSION UNSET query_tag;
```

Managed (or Reader) Accounts

Managed accounts enable providers to share data with non-Snowflake customers as they are created, managed, and owned by the provider account. In addition to creating managed accounts for our client use, managed accounts are also useful for local testing. You can find more information on managed accounts at `https://docs.snowflake.com/en/user-guide/data-sharing-reader-create`.

Figure 10-2 shows the relationship between Snowflake-supported containers used to implement data sharing. Note the tight coupling between a primary account and a managed account.

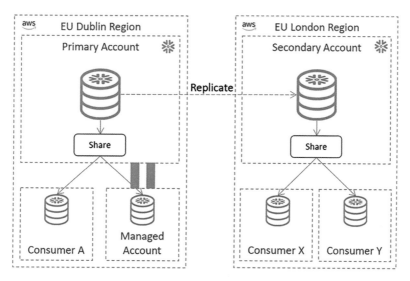

Figure 10-2. *Managed account in context*

Managed accounts use the *same* credit allocation as the creating primary account. Provisioning managed accounts may lead to uncontrolled credit consumption. If managed accounts are provisioned, warehouses in the managed account can consume unlimited credits from the providers budget; therefore, a resource monitor to limit usage should be set up. You can find more information on resource monitors at `https://docs.snowflake.com/en/sql-reference/sql/create-resource-monitor`.

I do not advise the creation of managed accounts but instead prefer each client to develop their own relationship with Snowflake Inc. However, managed accounts can be useful when testing shares as this approach removes the need to spin up a second Snowflake account.

Replication

Replication data volumes are impacted by your approach to data ingestion and curation. As covered in Chapter 8, parallelization has the potential to minimize replicated data through minimizing the number of micro-partitions changed for a given object. You will observe that data replication costs can far exceed curation costs; therefore, you must retain a tight focus on all parts of the data distribution.

Chapter 9 discussed how entitlements can be implemented, offering two different implementation patterns. The "all-or-nothing" approach leads to a high probability of data being replicated but with significant portions unused by clients. The bespoke client-centric approach of prefiltering data prior to presentation for client consumption may reduce replicated data sets, but at additional provider curation and refresh cost.

Snowflake replication costs are always paid by the data provider regardless of the data transfer mechanism.

From a producer perspective, you should always be mindful of your client consumption costs. From a consumer perspective, you should insist your consumption costs are minimized. Producers should facilitate their consumers by enabling multiple accounts to ingest the *same* share. Where data products are subject to user license limits, I suggest enabling multiple accounts to consume the same share offers additional sales opportunities due to higher usage and increased data product license requirements.

Multiplatform Distribution

Snowflake does not exist in isolation as the sole data marketplace. Figure 10-3 illustrates one possible ecosystem where Snowflake is used to curate and master data products and then distribute to multiple distribution venues.

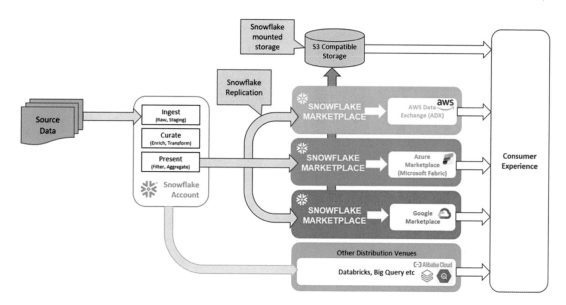

Figure 10-3. *Multiplatform data publishing*

Figure 10-3 does not cover all distribution venue possibilities; other cloud distribution venues are available.

Consumption Monitoring

Consumption metrics for published data will prove very useful to our marketing and sales colleagues. I suggest that collating all available consumption metrics into a consolidated reporting account for centralized reporting is sensible, as shown in Figure 10-4.

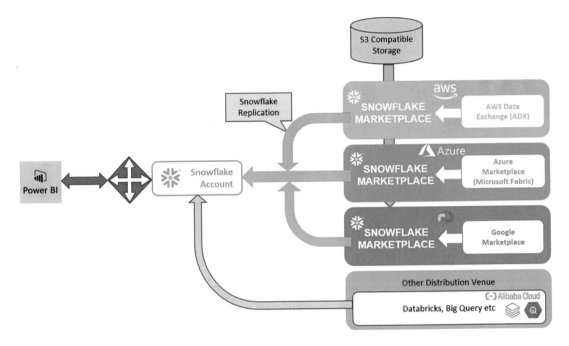

Figure 10-4. *Multiplatform metric ingestion*

Available consumption metrics will not include detailed SQL statements as these are likely to contain consumer intellectual property. Snowflake explicitly prevents producers from accessing their client SQL statements. Nonetheless, collated consumer metrics offer a degree of insight. Note that each platform's metrics may require conforming to a certain format for reporting purposes.

Optimizing Consumption

When developing applications, you will usually focus on achieving both system performance and cost to deliver data products. Usually this is a linear process deliver to either "like-for-like" capability when porting from an existing application or to build a minimum viable product for delivery to clients. Rarely do developers consider how clients will interact with a data product.

When ingesting data and combining intellectual property to curate data products, you must also consider adding features to facilitate client consumption of your data products. Some attributes lend themselves to improving query performance at little to no cost.

Consider date ranges. Many data feeds contain date attributes, and some are suitable for summarizing into an additional attribute containing YYYYMM only. By adding a new month-based attribute, you will enable filtering by month, which may facilitate micro-partition pruning.

Design for consumption. Ingestion (usually) happens once, whereas consumption happens many times for the same data set.

Benchmark CSP Performance

Most organizations have strategic platforms and commercial arrangements with one, two, or all three CSPs. Where choice exists, perform benchmark testing to identify the best-performing CSP for identical workloads. We are aware of performance differences in comparable virtual environments across different CSPs.

While not broken down by CSP, Snowflake publishes a performance index showing workload performance improvements over time for which further information can be found at `https://www.snowflake.com/en/data-cloud/pricing/performance-index/`.

Query Performance

Identifying query performance can be performed using several out-of-the-box tools; we highlight some available options next.

Figure 10-5 shows the Snowsight tools referenced in this section.

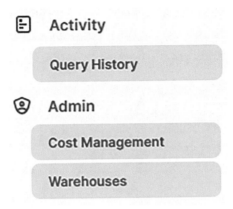

Figure 10-5. *Snowsight performance-related tooling*

The tools shown in Figure 10-5 provide a quick way to identify recent performance hot spots but do not provide wider contextual information required for a thorough investigation.

Snowsight is constantly evolving with new functionality and screens. We recommend periodic review of Snowsight capabilities; in fact, while writing this book new screens became available.

Warehouse Monitor

Snowsight offers a Warehouse screen accessible from the Admin link on the left side of the browser. Select Warehouses where summary information for the previous two weeks is displayed (shorter time periods can be displayed).

In the Query History section, the Status drop-down list provides several options, two of which are Queued and Blocked. Both options provide immediate visibility of recent issues with corresponding queries available for selection.

Cost Management Screen

Snowsight offers a Cost Management screen accessible from the Admin link on the left side of the browser. Select Cost Management and then Account Overview, where summary information about costs and the most expensive queries are displayed. At the time of writing, Account Overview is in public preview, and the capability will increase over time.

Query History

Snowsight offers a Query History screen accessible from the Monitoring link on the left side of the browser. Select Query History where the summary information for the previous 14 days is displayed; shorter time periods can be displayed along with a Custom option where a user-defined time period can be selected.

Use the DURATION column to order results by execution runtime.

Query Profile

Query profiles are useful for providing a visual indication of query execution operations and for identifying the following:

- How the query is physically executed

- Warehouse size used for execution

- Query execution order

- Ordered list of operation costs

- Spills to disk and out-of-memory (OOM) errors

- Cache reuse

- Micro-partition pruning

I discussed each aspect in detail in Chapter 3, noting preference for the following:

- Small build-side tables

- Large probe-side tables

- Right deep tree joins

Recognizing "good" patterns is important when tuning Snowflake code; the query profile tree and profile overview provide invaluable tools for performance tuning queries.

Figure 10-6 shows an example query profile reused from Chapter 3; note that the color coding in the PDF version is mine.

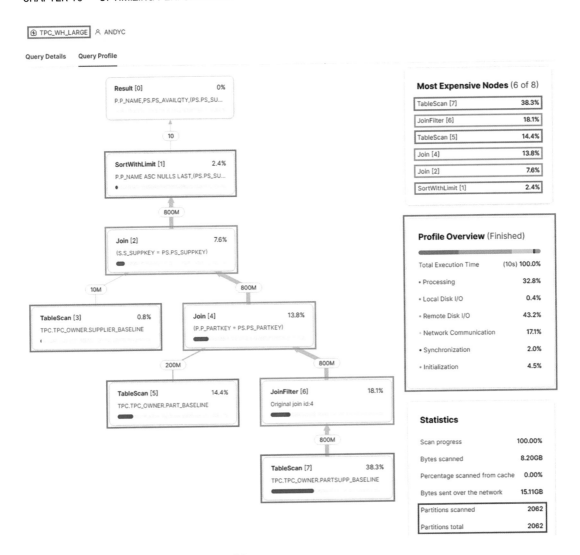

Figure 10-6. *Sample query profile*

Explain Plan

Evaluating queries before execution is another invaluable tool when performance tuning code. An explain plan shows the sequential steps performed by the optimizer when running the SQL statement. You can use an explain plan to evaluate the compile-time query plan, but note that an explain plan cannot expose runtime optimizations applied by the execution engine. These optimizations may occur due to prefiltered data from earlier processing, data distribution patterns, and automatic data skew optimizations.

You must regard an explain plan as a window into the optimization process just prior to the point of actual execution where further optimizations may occur.

Explaining SQL statements is a metadata operation; therefore, no warehouse is required; zero runtime cost are incurred though cloud service costs accrue.

In addition to the immediate knowledge gained for SQL statement execution, other use cases illustrate the value of explain plan.

- Delivering explain plan output as part of the software delivery life cycle provides a baseline for reference when the application has matured.

- Using an explain plan is good practice, instills discipline, and catches preventable issues early.

- Tabular format allows programmatic identification of issues.

To illustrate how EXPLAIN works, let's reuse the earlier known-good example query referencing v_supplier_part. In this example, I will request TABULAR output but might instead prefer JSON output:

```
EXPLAIN USING TABULAR
SELECT <attributes>
FROM   <table>
WHERE  <predicates>
ORDER BY <ordering>;
```

Figure 10-7 reuses the sample explain plan output from Chapter 3; note that further information is available if you scroll off to the right of the screen (not shown).

step	id	parent	operation	objects	alias	expressions
null	null	null	GlobalStats	null	null	null
1	0	null	Result	null	null	P.P_NAME, PS.PS_AVAILQTY, (PS.PS_SU
1	1	0	SortWithLimit	null	null	sortKey: [P.P_NAME ASC NULLS LAST,
1	2	1	InnerJoin	null	null	joinKey: (S.S_SUPPKEY = PS.PS_SUPPK
1	3	2	TableScan	TPC.TPC_OWNER.SUPPLIER_BASELINE	S	S_SUPPKEY, S_NAME, S_ACCTBAL
1	4	2	InnerJoin	null	null	joinKey: (P.P_PARTKEY = PS.PS_PARTKE
1	5	4	TableScan	TPC.TPC_OWNER.PART_BASELINE	P	P_PARTKEY, P_NAME
1	6	4	JoinFilter	null	null	joinKey: (S.S_SUPPKEY = PS.PS_SUPPK
1	7	6	TableScan	TPC.TPC_OWNER.PARTSUPP_BASELINE	PS	PS_PARTKEY, PS_SUPPKEY, PS_AVAILQ

Figure 10-7. *Sample explain plan output*

An explain plan also identifies micro-partition pruning through the GlobalStats and TableScan operators. You can also see the table aliases declared under the alias column; I will discuss aliases later. Figure 10-8 shows the effect of micro-partition pruning for the table PART_BASELINE. Note that partitionsAssigned is not an absolute value subject to later execution optimization.

operation	objects	alias	expressions	partitionsTotal	partitionsAssigned
GlobalStats	null	null	null	2060	1754
Result	null	null	P.P_NAME, PS.	null	null
Sort	null	null	P.P_NAME ASC	null	null
InnerJoin	null	null	joinKey: (PS.PS	null	null
InnerJoin	null	null	joinKey: (P.P_P.	null	null
Filter	null	null	P.P_NAME = 'a	null	null
TableScan	TPC.TPC_OWNER.PART_BASELINE	P	P_PARTKEY, P_	309	3

Figure 10-8. *Explain plan micro-partition pruning*

GET_QUERY_OPERATOR_STATS

GET_QUERY_OPERATOR_STATS returns query operator information for completed queries. GET_QUERY_OPERATOR_STATS is limited to queries executed in the past 14 days, which sets the maximum timeframe for how frequently any automated monitoring solution can run without missing information.

The general form of GET_QUERY_OPERATOR_STATS is as follows:

```
SELECT  <attributes>
FROM    TABLE ( get_query_operator_stats(<your value here>))
WHERE   <predicates>
ORDER BY <ordering>;
```

GET_QUERY_OPERATOR_STATS accepts a single value, which must be one of the following:

- The value returned by last_query_id()

- A session variable containing a valid query_id

- A string literal set to valid query_id

You might also use GET_QUERY_OPERATOR_STATS to investigate the most recently executed SQL statement:

```
SELECT *
FROM    TABLE ( get_query_operator_stats(last_query_id()));
```

Refer to Chapter 4 for further information on using GET_QUERY_OPERATOR_STATS. You can find more information about common query problems identified by GET_QUERY_OPERATOR_STATS at https://docs.snowflake.com/en/user-guide/ui-snowsight-activity#common-query-problems-identified-by-query-profile.

Optimizing Code

In this section, you will consider how to optimize your code to match Snowflake's "best practices" to optimize both costs and query performance. Some of the identified actions are zero risk and immediate benefits; others are more invasive.

You are responsible for ensuring the quality of your submitted SQL statements. Keep It Simple, Stupid.

Time Travel Setting

Incorrectly setting Time Travel to retain data for longer than required incurs additional storage costs. In Chapter 4, I discussed various considerations that may facilitate a reduction in your Time Travel setting for each object.

- Where ingested data can easily be reloaded, choose either temporary or transient tables.

- Where processed data is subject to high-frequency, low-volume DML activity, set Time Travel as low as acceptable.

- Build intermediate data sets into temporary tables before loading into core tables.

- Parallelize high-frequency, low-volume data loads to reduce micro-partition churn.

- Adopt an insert-only design pattern such as Data Vault 2.0.

- Where consumed data is periodically recreated, choose transient tables.

- For large tables, implement optimal clustering keys to match the most common data access paths.

Use Clones

When testing, you will often require either a baseline data set or a known configuration to reset at a known point in time. I will discuss cloned objects in detail in Chapter 4.

The overuse of clones is guaranteed to increase storage costs and, under certain circumstances, may contribute to metadata queries running slowly.

Remove all redundant cloned objects at the earliest possible opportunity.

Warehouse AUTO_SUSPEND

The minimum runtime for a warehouse is 60 seconds. You must ensure your warehouses cease execution after 60 seconds; this occurs only where there is no load on the warehouse.

```
SHOW warehouses;
ALTER WAREHOUSE compute_wh SET auto_suspend = 60;
```

Unless there are compelling reasons to retain an active, running warehouse, repeat for every warehouse in the account.

Warehouse Size

Don't be frightened of using a Large or bigger warehouse. They can be more time and cost effective than using a smaller warehouse. Note that failure to set the warehouse size correctly will result in excessive consumption charges.

Conversely, and in line with the best practices outlined in this book, if your code has been properly tuned, then you may be able to reduce your warehouse size while achieving the same or better process runtimes. Reducing warehouse size by one size *halves* the runtime cost.

Always seek to parallelize processes to concurrently use all the available warehouse processing units.

Warehouse scaling must also be considered; I will discuss warehouse scaling in Chapter 6.

Warehouse Usage

There are many considerations when optimally sizing and using your warehouses. Some of these are as follows:

- Is the workload consistent with historical "steady-state" workloads?

- How many concurrent workloads are running against the warehouse?

- What is your warehouse concurrency set to?

- Are workloads queueing?

- Is warehouse clustering enabled and, if so, to what degree?

- For each workload, are any workloads spilling to disk?

- Is object locking evident?

- Does the warehouse run too frequently?

- Are too many warehouses of same size declared?

- Is there low warehouse cache reuse?

- Is there an incorrect `auto_suspend` setting?

- Is there an artificial warehouse size constraint imposed?

- Are files correctly sized for ingestion?

- Is the warehouse correctly sized for the workload?

- Is serial or parallel logging implemented?

- Are warehouse scaling policies appropriate for the operating environment?

- Are warehouse modes correct for the expected workload?

To reduce the number of warehouses, use query tags when consolidating workloads onto warehouses.

Warehouse Scaling Policy

By default, multicluster warehouses are created with the Standard scaling policy. While appropriate for production environments, most nonproduction environment warehouses will benefit from using the economy scaling policy to achieve a better balance of speed and performance. With the economy scaling policy, nonproduction environments can tolerate a higher degree of queueing to ensure greater warehouse processing unit utilization.

```
ALTER WAREHOUSE <warehouse_name> SET SCALING_POLICY = 'ECONOMY';
```

You can find more information on warehouse scaling policies at `https://docs.snowflake.com/en/user-guide/warehouses-multicluster#setting-the-scaling-policy-for-a-multi-cluster-warehouse`.

Warehouse Mode

Warehouses operating in auto-scale mode are identified where the maximum and minimum number of clusters have *different* values. In this scenario, a single cluster is started at warehouse instantiation with further clusters starting (subject to maximum clusters setting) according to workload. Auto-scale is the most common warehouse mode and is useful for varying workloads.

```
ALTER WAREHOUSE <warehouse_name> SET MIN_CLUSTER_COUNT = 1;
ALTER WAREHOUSE <warehouse_name> SET MAX_CLUSTER_COUNT = 4;
```

Warehouses operating in maximized mode are identified where the maximum and minimum number of clusters are the *same* value. In this scenario, all clusters are started at warehouse instantiation and may be useful for known, static workloads.

```
ALTER WAREHOUSE <warehouse_name> SET MIN_CLUSTER_COUNT = 2;
ALTER WAREHOUSE <warehouse_name> SET MAX_CLUSTER_COUNT = 2;
```

You can find more information on warehouse maximized mode at `https://docs.snowflake.com/en/user-guide/warehouses-multicluster#maximized-vs-auto-scale`.

Bind Variables

Bind variables enable query reuse. The first time a SQL statement is seen by the optimizer, a hard-parse is performed, all subsequent query submissions reuse the original execution plan and substitute values for the bound variables.

Overall query performance suffers where SQL statements are always hard-parsed, the optimizer cannot re-use queries. Bind variables are always considered best practice where the main body of a query remains static. Implementation costs for implementing bind variables result in both lower execution costs and reduced code maintenance overheads.

Eliminate SELECT *

Snowflake only access those explicitly named attributes from base tables. SELECT * is an obvious candidate for removal and replacement with explicitly named attributes. In real-world testing we observe performance improvements by removing SELECT *.

Replace SELECT * with only those attributes required to satisfy the query. Do not add extraneous or unused attributes as these result in extra workload.

SELECT * with UNION (not UNION ALL) results in a full table scan for each side of the UNION.

Eliminate DISTINCT

I occasionally observe the use of the DISTINCT clause to enforce uniqueness. Closer examination often reveals a missing join condition from the query predicates or poor data model implementation.

Identify and remove all DISTINCT clauses wherever possible, and use the previously covered query performance tools to aid investigation.

Examine Common Table Expressions (CTEs)

Where a CTE is referenced more than once in the same SQL statement, you may find attribute pruning is disabled.

I recommend the use of CTEs to abstract complex logic and simplify code, but not in all situations.

Please refer to the corresponding section in Chapter 3 for further details.

Window Functions

QUALIFY provides the same functionality for window functions as HAVING does for aggregate GROUP BY functions. QUALIFY may reduce memory usage by limiting results; see https://docs.snowflake.com/en/sql-reference/constructs/qualify.

Use the *same* keys for PARTITION BY and ORDER BY clauses. Using different keys will result in a performance penalty.

Implement a single consistent windowing pattern for multiple analytic function calls in the same SQL statement.

Always implement a PARTITION BY clause in a window function regardless of whether the query is successfully processed. Where no obvious partitioning pattern matches the requirement, use either PARTITION BY NULL or PARTITION BY 1.

Returned Query Attributes

Snowflake prefers fewer attributes to be returned from individual queries and suggests using 100 or fewer attributes (according to Jiaqi Yan, principal software engineer, and Minzhen Yang, principal engineer and tech lead, at Snowflake Inc).

Micro-partitions underpin every SQL statement. You must consider how data is stored and maintained in storage when making query decisions. You should optimize your physical base table storage for query performance. Performance may be improved by separating VARIANT attributes into a separate table where most queries do not reference the VARIANT attributes.

Reduce Nested Views

Wherever possible, nested views should be removed from queries. In general, query optimizers must resolve each nested view before resolving the main query. You can see how nesting views may lead to performance issues both in terms of increased query compilation time and execution time.

Nested views are often difficult to debug as they act as a translation, filter, summary, or aggregate layer between source objects and consuming objects or queries. Temporarily replacing an intermediary view with a table of the same name and contents helps resolve issues as the query profile will be far simpler. Better still, replace nested views with more elegant SQL encompassing view functionality.

I discuss how to identify object types later in this chapter as the naming convention alone cannot be relied upon.

Replace Subqueries

Snowflake may not always optimize subqueries to dynamically prune micro-partitions, and rewriting a query may not always convert a subquery to a join. Instead, convert subqueries to direct joins or CTEs where appropriate. There is usually no benefit to ordering data in a subquery or CTE except to obtain an intermediate top 'n' rowset.

As always, test and then retest to prove that the changes are effective.

Optimization Focus

When optimizing SQL statements, the most impact will be realized by optimizing the number of rows returned. In order of preference, your approach should focus on the following:

- Reducing the number of objects accessed

- Ensuring join conditions are complete and correct

- Making sure filter criteria are sufficiently selective and match any defined clustering keys

- Minimizing aggregations and aggregate filters

- Reducing analytic operations

- Removing ordering and record limits

Optimize INSERTs

Occasionally you will encounter individual INSERT statements that may be consolidated into multirow INSERT statements.

A multirow INSERT statement is considerably more efficient than many individual INSERT statements because of the immutable nature of micro-partitions. Every individual INSERT statement causes a new micro-partition write *for each row*, whereas a multirow INSERT causes fewer micro-partition writes.

UNION or UNION ALL

UNION forces a SORT, whereas UNION ALL does not enforce a SORT. Replace UNION with UNION ALL where appropriate.

Joins

Joins can be improved by rewriting code for efficiency; I explain some of the more common issues encountered with joins next.

Remove Disjunctive Joins

The Snowflake optimizer prefers conjunctive (additive) joins; these are predicates with AND operators. Predicates with OR operators are disjunctive (subtractive) joins that are known to affect performance. Disjunctive joins should be rewritten using UNION/UNION ALL to improve performance.

Missing Joins

Identify missing join criteria as this is the most likely root cause. Note that missing composite key attributes are far harder to identify than single attribute primary key/ foreign key relationships. A general rule of thumb is that the number of AND join conditions should always be equal to the number of tables minus 1. This works for many scenarios.

Type Casting

Multiple layers of type casting on join keys prevents static micro-partition pruning at compile time. Consider adding typecast attributes at table creation time for population by ingestion or curation processes. By way of example, add an "YYYYMM" attribute where a DATE attribute would normally be used. The additional low-cardinality attribute may be a suitable candidate for clustering key definition to enable more efficient pruning.

Optimizing Joins

Numeric data type joins are the fastest of all. I prefer sequence generated surrogate primary keys over natural or composite keys for all tables along with declared referential integrity. Numeric data types are also preferred for clustering keys where the number range is low cardinality. Sequences do not make good candidates for leading attributes in clustering keys.

Dates and timestamps are stored internally in numeric format and therefore are good candidates for both join conditions and clustering keys. I prefer to reduce the cardinality of both date and timestamp attributes when used as leading attributes in clustering keys; I suggest reducing cardinality to the year/month (YYYYMM) format to improve pruning.

Cardinality can be determined from both metadata and Snowflake-supplied nondeterministic estimation functions such as HYPERLOGLOG for which further information can be found here at https://docs.snowflake.com/en/sql-reference/functions/hll. Bitmaps can also be used to improve performance; see https://docs.snowflake.com/en/user-guide/querying-bitmaps-for-distinct-counts.

Remove type casting from join key attributes; instead, pre-type cast into new attributes at the point of data ingestion.

Table Join Order

Table join order on SQL statements can be significant. Start with the smallest cardinality tables first to eliminate the greatest number of micro-partitions early in the query optimization stage. Also check that the query filter criteria are sufficiently selective to improve micro-partition pruning.

Simplify Logic

Wherever possible, remove aggregations and summaries from join, group by, and order by operations as cardinality estimates suffer. Instead, create materialized views to pre-aggregate and summarize attributes.

Reduce the number of levels navigated to resolve query result sets.

Missing Referential Integrity

Referential integrity may be retrospectively applied by using an ALTER TABLE statement; see https://docs.snowflake.com/en/sql-reference/sql/create-table-constraint.

Missing Aliases

At all times you should remove metadata lookups by adding aliases to all referenced objects. Adding meaningful aliases aids readability and code maintenance too.

Temporary Tables

Temporary tables are restricted to the local session and are removed when the session closes. Temporary should be regarded as an interim, nonpersistent step in a process, recognizing the contents cannot be inspected after the session closes.

Temporary tables have these characteristics:

- Avoid name conflicts for temporary tables with base tables. If common, then the temporary table will be referenced in the statement.

- Separate micro-partition metadata and statistics maintained throughout their life cycle.

- Materialize intermediate data sets for use in the session.

- Statistics inform the optimizer decision-making process.

- Work with `EXPLAIN PLAN`.

Set LIMIT

While developing applications, you may need to examine a small data sample to identify filters and test your code functions correctly. Restricting sample data is readily achieved by adding a `LIMIT` clause to your SQL statements.

Skewed Data

Performance tuning is not a once-off activity. Data profiles change over time, and `INSERT`, `UPDATE`, and `DELETE` operations can cause skewed data where the distribution of data in a table or database becomes increasingly imbalanced or uneven. The impact of data skew over time can be significant, particularly when it comes to query performance.

Hash partition joins are used to join large tables, and skewed data may result in warehouse processing unit overload. Skewed data may be handled more efficiently by parallelizing data operations, as explained in Chapter 8.

Skewed data may impact clustering. I will discuss clustering keys next.

Ineffective Pruning

Micro-partition pruning is dependent upon several factors including the following:

- Use of appropriate filter attribute(s)

- Simple filter expressions, i.e., no operators applied.

- For unclustered tables, filter attributes matching natural clustering order of ingested data

Where a table has an explicitly declared clustering key and our SQL statement predicates do not match the clustering key declaration, a materialized view may provide an optimal search path noting the additional storage and serverless compute required for maintenance.

Fully Sorted Table

When rebuilding tables, you may prefer to initialize the data by pre-ordering using the explicit ORDER BY syntax. This approach works for both CTAS and INSERT OVERWRITE. Pre-seeding data in an ordered manner may facilitate later creation of a clustering key.

The initial creation of micro-partitions for pre-ordered data will be faster than later relying upon the automated clustering service where a clustering key is defined. However, spills to disk and OOMs may occur. Subsequent DML operations may result in skewed data.

Clustering Keys

Clustering keys should not be your first consideration when performance tuning code. Ensure all other options have been tried first.

Clustering key maintenance is impacted by high-velocity, low-volume Data Manipulation Language (DML) operations where the asynchronous Automatic Clustering service may not cope with the speed of change.

Clustering keys prefer low-cardinality leading attributes to maximize pruning, and clustering keys should be limited to three or four attributes only. For String datatypes, only the first five to six characters should be used to minimize cardinality; otherwise, numeric datatypes with low cardinality are preferred.

Clustering keys support micro-partition pruning. Optimize key attributes for maximum pruning effectiveness.

Most use cases do not require a clustering key. Snowflake suggests tables of 1TB or greater should be considered as candidates for clustering key declarations. For those use cases where a clustering key is defined, you must understand the attribute order and cardinality, preferring low-cardinality attributes first.

Clustering keys alone are not a silver bullet. They are part of an overall performance tuning strategy.

Operational considerations for clustering keys management include the following:

- Clustering keys should be defined only for high-frequency queries with matching query predicates.

- Clustering key maintenance may not happen concurrent with DML operation completion; maintenance requires a finite time to complete.

- Frequent DML operations may result in costly reclustering operations and, in worst-case scenarios, constant micro-partition "churn."

- Clustering is most cost effective for low-volume DML and high-volume query operations.

- Reclustering invalidates cached results.

For those inclined to dive deeper into clustering, the Snowflake founders' patent can be found at `https://www.freepatentsonline.com/y2018/0068008.html`.

Please refer to Chapter 5 for a thorough investigation of clustering keys, paying particular attention to clustering width and clustering depth.

Introspection Calls

In Snowflake, an introspection call is a SQL statement used to interrogate the Account Usage Store or information schema of a particular database to identify metadata for objects, columns, and their attributes.

Performance issues with introspection calls may be caused by the following:

- Unexpected or unpredictable changes made to object definitions causing ad hoc metadata changes

- High numbers of nested roles or masking policies

- High number of redundant cloned objects

As suggested by Nadir Doctor, queries using Account Usage Store can be significantly improved via referencing a base local target table, created via a CTAS operation to contain a backup of data from a source view.

File Size Optimization

The recommended file size for Snowpipe and COPY commands is 100MB to 250MB compressed. Ingesting smaller files leads to both increased cost and longer warehouse runtimes.

Check All Tasks

Tasks may fail for a variety of reasons, many of which can be diagnosed from information found here: `https://docs.snowflake.com/en/user-guide/tasks-ts`. Tasks may auto-suspend according to the value of the parameter `SUSPEND_TASK_AFTER_NUM_FAILURES` for which further information can be found here: `https://docs.snowflake.com/en/sql-reference/parameters#suspend-task-after-num-failures`.

For tasks dependent upon streams, when the stream goes stale, the task will fail. The parameter `MAX_DATA_EXTENSION_TIME_IN_DAYS` can be set independent of the parent table; see `https://docs.snowflake.com/en/sql-reference/parameters#max-data-extension-time-in-days`.

Periodically check tasks to see if any can be disabled or, better still, removed.

Check task start times to determine whether tasks can be retimed and/or consolidated into fewer warehouses to parallelize processing.

Session Settings

This section illustrates some useful SQL statements to aid your testing.

Statement Timeout

While testing code, you may prefer to set `statement_timeout_in_seconds` in the current session to avoid overspend. In this example, I set the timeout to 600 seconds (10 minutes), but note the smaller of the user or warehouse setting applies:

```
ALTER SESSION SET statement_timeout_in_seconds = 600;
```

Statement Queueing

Except when set to zero, Snowflake automatically cancels individual queries queued in excess of `statement_queued_timeout_in_seconds`.

Task Timeout

An individual task invocation will run for `user_task_timeout_ms` before being cancelled by Snowflake.

Concurrency

The number of concurrent processes executed by a warehouse can be set by `max_concurrency_level`.

Execution Context

The current session execution context can be derived by this query:

```
SELECT current_account(),
       current_user(),
       current_role(),
       current_warehouse(),
       current_database(),
       current_schema();
```

Clearing Warehouse Cache

To ensure raw performance figures are not skewed by the cache when testing or investigating, ignore cached results causing every SQL statement to be executed.

```
ALTER SESSION SET use_cached_result = FALSE;
```

The warehouse declaration does not clear out the warehouse cache, so you must suspend and restart the warehouse, which also aborts all the active SQL statements.

```
USE WAREHOUSE IDENTIFIER ( $tpc_warehouse_xs );

ALTER WAREHOUSE IDENTIFIER ( $tpc_warehouse_xs ) SUSPEND;

ALTER WAREHOUSE IDENTIFIER ( $tpc_warehouse_xs ) RESUME;
```

Micro-partition reclustering or consolidation causes cached result sets to be invalidated preventing reuse.

Referenced Objects

When tuning SQL, you must understand the type of object referenced. Object naming conventions alone are no guarantee of the underlying object type; you must explicitly know whether your code addresses tables, external tables, hybrid tables, dynamic tables, views, or materialized views.

For dynamic tables, views, and materialized views, the distinction is important. All of these objects contain stored queries that must be executed to return a summary data set before being joined to other tables and views. You may also encounter the following:

- Latency for dynamic tables due to underlying data changes not being immediately reflected into the object

- Performance penalty for materialized views where Snowflake either updates the materialized view or uses the up-to-date portions of the materialized view and retrieves any required newer data from the base table

- Views and secure views exhibiting different performance characteristics

- Increased costs for maintaining dynamic tables and materialized views for frequent underlying data changes

You can find more information on views, secure views, and materialized views at `https://www.linkedin.com/pulse/materialized-view-vs-secure-regular-minzhen-yang/`.

Nesting logic in views is a common way to abstract (or hide) complexity, and experience suggests performance issues may be buried inside views.

Identifying Object Types

When performance tuning SQL statements, you must identify all in-scope objects for the current role.

Let's start by setting the execution context.

```
SET tpc_owner_role   = 'tpc_owner_role';
SET tpc_warehouse_XS = 'tpc_wh_xsmall';
SET tpc_database     = 'tpc';
SET tpc_owner_schema = 'tpc.tpc_owner';
```

```
USE ROLE      IDENTIFIER ( $tpc_owner_role   );
USE DATABASE  IDENTIFIER ( $tpc_database     );
USE SCHEMA    IDENTIFIER ( $tpc_owner_schema );
USE WAREHOUSE IDENTIFIER ( $tpc_warehouse_xs );
```

Somewhat surprisingly, Snowflake does not provide a single account_usage or information_schema view to identify all objects, their types, and their location. Investigating the available account_usage or information_schema views may return misleading results. For example, dynamic tables are referenced as type NULL.

SHOW always returns information for the current role.

At the time of writing, using the SHOW OBJECTS command also returns the incorrect object type of VIEW. Where you would expect to see MATERIALIZED VIEW, a change request with Snowflake has been raised to normalize behavior.

```
SHOW OBJECTS;
```

However, you find issuing SHOW command for each target object type works as expected. Note that each SHOW command returns differing attributes.

```
SHOW TABLES;
SHOW DYNAMIC TABLES;
SHOW EXTERNAL TABLES;
SHOW HYBRID TABLES;
SHOW VIEWS;
SHOW MATERIALIZED VIEWS;
```

Your objective is to extract a consistent form of metadata for each of the SHOW commands for consistent later use. This SQL statement works for the previous SHOW commands. Note that the object_type should be changed according to the desired object type. You can also filter out objects created using supplied Snowflake roles.

```
SELECT "database_name"||'.'||"schema_name"||'.'||"name" AS path_to_object,
       'TABLE | VIEW | etc' AS object_type,
       "owner",
       "owner_role_type"
```

```
FROM TABLE ( RESULT_SCAN ( last_query_id()))
WHERE   "owner" NOT IN ( 'ACCOUNTADMIN', 'SECURITYADMIN', 'SYSADMIN')
ORDER BY 1 ASC;
```

Identifying procedures and functions is largely similar in form to the tables and views shown earlier:

```
SHOW PROCEDURES;
SHOW FUNCTIONS;
```

As shown earlier, your objective is to extract a consistent form of metadata for each of the SHOW commands for consistent later use. This SQL statement works for the previous SHOW commands noting the object_type should be changed according to the desired object type. You can also filter out objects created without a schema; the schema_name is empty string.

```
SELECT "catalog_name"||'.'||"schema_name"||'.'||"name" AS path_to_object,
       'PROCEDURE | FUNCTION' AS object_type
FROM TABLE ( RESULT_SCAN ( last_query_id()))
WHERE   "owner" NOT IN ( 'ACCOUNTADMIN', 'SECURITYADMIN', 'SYSADMIN')
AND     "schema_name" != ''
ORDER BY 1 ASC;
```

You may prefer to wrap all the previous SHOW commands and extend subsequent SQL statements to insert into tables. I suggest a stored procedure encapsulating logic into a single container for periodic reuse.

Identifying Object Dependencies

Object dependencies are created when a parent object references a child object. For example, a view is an example of a parent object with dependencies on those objects referenced in the view declaration. In this manner, you can observe object dependencies exist in hierarchical form, as shown in Figure 10-9.

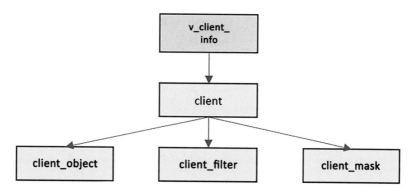

Figure 10-9. *View v_client_info object dependencies*

Object dependencies may exist across more than one database in an account. Snowflake maintains object dependency metadata in the Account Usage Store view object_dependencies. Note that latency of up to three hours may be experienced.

```
SELECT referenced_database||'.'||
       referenced_schema||'.'||
       referenced_object_name AS path_to_object,
       dependency_type
FROM   snowflake.account_usage.object_dependencies
WHERE  referencing_database   = 'TPC'
AND    referencing_schema     = 'TPC_OWNER'
AND    referencing_object_name = 'V_CLIENT_INFO'
ORDER BY 1;
```

A more sophisticated method for deriving object dependencies involves the use of a "tree walk" where a top-level object is named, and all child dependencies are resolved. Note the inclusion of both parent and child attributes.

```
SELECT referenced_database||'.'||
       referenced_schema||'.'||
       referenced_object_name  AS path_to_object,
       referencing_database||'.'||
       referencing_schema||'.'||
       referencing_object_name AS path_to_parent,
       dependency_type
```

```
FROM    snowflake.account_usage.object_dependencies
WHERE   referencing_database          = 'TPC'
AND     referencing_schema            = 'TPC_OWNER'
START WITH referencing_object_name = 'V_CLIENT_SQL_STATEMENT'
CONNECT BY referencing_object_name = PRIOR referenced_object_name
ORDER BY 1;
```

These approaches also work for materialized views but *do not work* for dynamic tables.

You can find more information on the object_dependencies view at https://docs.snowflake.com/en/sql-reference/account-usage/object_dependencies. You can find a fuller explanation of how object dependencies are tracked at https://docs.snowflake.com/en/user-guide/object-dependencies.

Identifying Constraints

Referential integrity may exist across more than one database in an account. Snowflake maintains referential integrity metadata in the Account Usage Store view table_constraints. Note that latency of up to two hours may be experienced.

```
SELECT table_catalog||'.'||
       table_schema||'.'||
       table_name       AS path_to_object,
       constraint_name,
       constraint_type
FROM    snowflake.account_usage.table_constraints
WHERE   constraint_catalog     = 'TPC'
AND     constraint_schema      = 'TPC_OWNER'
AND     deleted IS NULL
ORDER BY 1;
```

You can find more information on the table_constraints view at https://docs.snowflake.com/en/sql-reference/account-usage/table_constraints.

For primary keys, only the SHOW command proves useful. Note the addition of attributes that comprise the primary key.

```
SHOW PRIMARY KEYS IN SCHEMA;
```

This SQL statement works for the previous SHOW commands, and note the explicit ordering of path_to_obejct and key_sequence to ensure primary key attributes are displayed consecutively.

```
SELECT "database_name"||'.'||"schema_name"||'.'||"table_name" AS path_
to_object,
        "column_name",
        "key_sequence",
        "constraint_name"
FROM TABLE ( RESULT_SCAN ( last_query_id()))
ORDER BY path_to_object, "key_sequence";
```

You might also use the referential_constraints view though the results are less useful. I leave this to you for your further investigation. See https://docs.snowflake.com/en/sql-reference/info-schema/referential_constraints. A fuller explanation of how constraints are supported can be found at https://docs.snowflake.com/en/sql-reference/constraints-overview.

GET_DDL

As you have seen, Snowflake does not provide a simple method to identify objects related by referential integrity; you can only access related information, not the relationship information.

An alternative method for identifying referential integrity declarations is to examine each object definition. In this example, you use the get_ddl function to extract object metadata for the table client_mask.

```
SELECT get_ddl ( 'TABLE', 'TPC.TPC_OWNER.CLIENT_MASK', TRUE );
```

The returned string contains the following:

```
create or replace TABLE CLIENT_MASK (
    CLIENT_MASK_ID NUMBER(38,0) NOT NULL,
    CLIENT_ID NUMBER(38,0),
    MASK_NAME VARCHAR(255),
    MASK_ATTRIBUTE VARCHAR(255),
    MASK_VALUE VARCHAR(255),
```

```
    primary key (CLIENT_MASK_ID),
    foreign key (CLIENT_ID) references TPC.TPC_OWNER.CLIENT(CLIENT_ID)
);
```

In the get_ddl returned string, you see foreign key (CLIENT_ID) references TPC.TPC_OWNER.CLIENT(CLIENT_ID), which may be programmatically extracted for use.

While not directly relevant to identifying referential integrity, get_ddl may also be used to derive definitions for a wide variety of objects, in this example, for a view called v_client_sql_statement.

```
SELECT get_ddl ( 'VIEW', 'V_CLIENT_SQL_STATEMENT' );
```

You can find information on the get_ddl view at https://docs.snowflake.com/en/sql-reference/functions/get_ddl.

User Defined Objects

In this section I identify some performance limitations and restrictions for several user-defined objects including tables, views, materialized views, dynamic tables, procedures, and functions.

Tables

I highlighted the differences between transient and permanent tables earlier in this chapter, and as I am discussing user-defined objects here, I will repeat the advice for completeness.

I suggest ingestion raw or staging tables should use transient tables with Time Travel set to 0 as transient tables do not utilize Fail-Safe. You might also consider transient tables for frequently refreshed data generated for point-in-time reporting.

Permanent tables should be reserved for persistent storage where Time Travel is required. Note that the seven-day Fail-Safe period follows. You should optimize the default database Time Travel setting along with each table Time Travel setting according to tour use cases. Setting Time Travel to 90 days is often overkill, and a shorter time period is preferable particularly where high-velocity, low-volume DML operations cause significant micro-partition churn.

Most use cases do not require a clustering key. For those use cases where a clustering key is defined, you must understand the attribute order and cardinality.

Views and Dynamic Tables

Both views and dynamic tables share the common feature of facilitating data model denormalization by joining tables and applying filters and aggregates to provide a composite representation of the result set. While the delivery mechanisms differ insofar as a view is an abstracted query and a dynamic table is a periodically refreshed abstracted query, the underlying principles for deriving data are the same.

Both views and dynamic tables may suffer from the same performance impacting issues, I highlight this when optimizing code. I prefer to drill down into all child views, building knowledge from the ground up and then deciding upon an appropriate course of action. For complex relationships, I suggest an entity relationship diagram (ERD) will assist in resolving performance issues.

While there are no additional storage costs for views, you can incur additional storage costs along with serverless compute costs for provisioning dynamic tables. As with standard tables, high-velocity, low-volume DML operations cause significant micro-partition churn for dynamic tables.

Secure Views

Secure views prevent the view definition from being exposed to unauthorized users and prevent access to the underlying SQL query for all roles except the role that owns the secure view. In addition to the comments highlighted earlier when optimizing code, the high-security profile for secure views restricts optimization to a subset of data points available for optimizing normal views.

During query processing pushdown optimization, the query processor prefilters rows by dynamically pruning micro-partitions to improve performance and reduce memory consumption. With normal views, pushdown can allow confidential data to be exposed indirectly. You can find more information on pushdown at https://docs. snowflake.com/en/developer-guide/pushdown-optimization.

Secure views prevent the exposure of confidential information; by default, most pushdown optimizations are disabled. These operations prevent pushdown, and there may be more:

- Arithmetic operations in query `WHERE` clauses

- `UNION` operations

- Scalar functions that take a row or value and return a single value

As with all SQL statements, avoid complexity, and simplify code wherever possible. There are no additional storage costs associated with secure views.

You can find more information on secure views at `https://docs.snowflake.com/en/user-guide/views-secure`.

Materialized Views

As discussed in Chapter 3, a materialized view can be declared only on a single table and is a way either to declare alternative clustering keys on a base table or to summarize or aggregate data. Using materialized views facilitates micro-partition pruning, as discussed in Chapter 4.

Materialized views incur maintenance, runtime, and storage costs. Before implementing and using materialized views, a balance must be struck to ensure optimally cost-effective solutions are developed and delivered. You can incur additional storage costs along with serverless compute costs for provisioning materialized views. As with standard tables, high-velocity, low-volume DML operations cause significant micro-partition churn for materialized views.

You can find information on materialized views at `https://docs.snowflake.com/en/user-guide/views-materialized`. I also found this article on the different types of views by Minzhen Yang useful: `https://www.linkedin.com/pulse/materialized-view-vs-secure-regular-minzhen-yang/`.

User-Defined Functions (UDFs)

In similar manner to views, UDFs provide the capability to implement bespoke functionality callable using standard SQL. The key difference between a view and a UDF is the degree of complexity that can be accomplished. UDFs offer a far wider range of programming options than SQL; UDFs can be implemented using Java, JavaScript, Python, Scala, and SQL.

Whenever you encounter UDFs embedded in SQL statements, the root cause is typically to abstract very complex logic to return a readily understood answer compatible with the calling SQL query body. UDFs are called for *every row* in the calling SQL body and therefore often result in performance bottlenecks.

Wherever possible, I prefer to remove inline UDFs and instead resolve complex logic using standard SQL. This approach does not suit all use cases.

You can find more information on UDFs at `https://docs.snowflake.com/en/developer-guide/udf/udf-overview`.

Identifying Issues

Previous sections in this chapter have addressed how to design for performance and remediate SQL statements to improve performance. In this section, I focus on identifying SQL statements by investigating the `information_schema.query_history` view for metrics. Note the 14-day limit on data retention. Alternatively, you may prefer to use the corresponding Account Usage Store view, which both retains information for 1 year and has up to 45 minutes latency.

Warehouse Queueing

Queueing is identified where `queued_overload_time` represents the amount of time the query waits before execution commences. The following query is offered as a starting point for your investigation. Chapter 6 contains a thorough investigation of warehouse queueing from which these queries are derived; please refer to Chapter 6 for a fuller explanation.

```
SELECT query_type,
       query_id,
       query_text,
       role_name,
       warehouse_name,
       queued_overload_time,
       execution_time
```

```
FROM    TABLE ( information_schema.query_history())
WHERE   queued_overload_time > 0
AND     execution_time       > 0;
```

You can find information on queueing at `https://community.snowflake.com/s/article/Understanding-Queuing`.

Warehouse Workload

Identifying workload peaks and troughs may provide the means to balance your workloads throughout the day. For example, housekeeping processes and generating summaries can often be moved to quiet times or parallelized to consume more processing units from a running but under-utilized warehouse. Please refer to Chapter 6 for a fuller explanation.

```
SELECT warehouse_name,
       start_time,
       end_time,
       query_id,
       query_text,
       total_elapsed_time / 1000  AS total_elapsed_time_in_secs,
       transaction_blocked_time,
       DATE_PART (         'YYYY', start_time )||
       LPAD ( DATE_PART ( 'MM',   start_time ), 2, '0' )||
       LPAD ( DATE_PART ( 'DD',   start_time ), 2, '0' )||'_'||
       LPAD ( DATE_PART ( 'HOUR', start_time ), 2, '0' )
                                   AS date_time
FROM    snowflake.account_usage.query_history
WHERE   execution_time <> 0
ORDER BY warehouse_name,
         start_time DESC;
```

The previous query provides a high-level view of activity only.

Blocked Transactions

Blocked transactions are those DML operations waiting for an object lock before completing. I previously discussed how multiple concurrent processes logging into a single table will serialize processing as each process must acquire a table lock before completing their DML operation. The Account Usage Store `query_history` table provides information on blocked transactions, as shown next:

```
SELECT query_type,
       query_id,
       query_text,
       total_elapsed_time,
       transaction_blocked_time
FROM    TABLE ( information_schema.query_history())
WHERE   transaction_blocked_time > 0
ORDER BY transaction_blocked_time DESC;
```

Blocked transactions may time out waiting for a lock, and you may set a session variable called `LOCK_TIMEOUT` to adjust the default from 12 hours to a more suitable value, in this example, an hour.

```
ALTER SESSION SET LOCK_TIMEOUT = 3600;
```

You can find more information on setting session values at `https://docs.snowflake.com/en/sql-reference/sql/alter-session`. I also found this article useful where `LOCK_TIMEOUT` is not honored by transactions: `https://community.snowflake.com/s/article/LOCK-TIMEOUT-not-honoured-by-transactions`.

You can find more information on blocked transactions at `https://docs.snowflake.com/en/sql-reference/transactions#label-analyzing-blocked-transactions`.

Join Explosion

Identifying join explosions is a two-step process. Here I present the two SQL statements required.

First, identify candidate long-running queries, noting that not all long-running queries will suffer from join explosion.

```
SELECT  query_id
FROM    TABLE ( information_schema.query_history())
WHERE   query_type     IN ( 'SELECT', 'CREATE_TABLE_AS_SELECT' )
AND     warehouse_name IS NOT NULL
AND     execution_status   = 'SUCCESS'
AND     bytes_scanned      > 0
AND     total_elapsed_time > 1000;
```

With the candidate query_id values identified, you can investigate each for CartesianJoin operations.

```
SELECT  operator_type,
        operator_id,
        operator_attributes,
        operator_statistics:output_rows / operator_statistics:input_rows AS
        row_multiple
FROM    TABLE ( get_query_operator_stats('<YOUR_QUERY_ID_HERE>'))
WHERE   operator_type = 'CartesianJoin';
```

Join explosions are usually caused by a missing join condition in the query predicates (WHERE clause).

Guidance on resolving join explosions is provided in Chapter 3. You can find more information on join explosion at https://community.snowflake.com/s/article/Recognizing-Row-Explosion and https://docs.snowflake.com/en/sql-reference/functions/get_query_operator_stats#identifying-exploding-join-operators.

Long Compilation Time

Long compilation time identifies records where the compilation time exceeds the execution time. Please refer to Chapter 3 for a full explanation.

```
SELECT  query_id,
        warehouse_name,
        warehouse_size,
        compilation_time,
```

```
       CASE execution_time
          WHEN 0 THEN 1
          ELSE execution_time
       END AS execution_time_1,
       compilation_time / execution_time_1
FROM   TABLE ( information_schema.query_history())
WHERE  ( compilation_time / execution_time_1 ) > 1
AND    warehouse_size IS NOT NULL;
```

Guidance on resolving long compilation time is provided in Chapter 3. You can find more information on long compilation time at https://community.snowflake.com/s/article/Understanding-Why-Compilation-Time-in-Snowflake-Can-Be-Higher-than-Execution-Time.

Long Execution Time

Long execution time occurs after a query has been compiled and relates to the physical amount of time required to return a result set. Please refer to Chapter 3 for a full explanation.

```
SELECT query_id,
       warehouse_name,
       warehouse_size,
       total_elapsed_time / 1000 AS query_execution_time_s
FROM   TABLE ( information_schema.query_history())
WHERE  warehouse_name IS NOT NULL
AND    execution_status   = 'SUCCESS'
AND    bytes_scanned       > 0
AND    total_elapsed_time > 1000;
```

Guidance on resolving long execution time is provided in Chapter 3. You can find more information on long execution time at https://docs.snowflake.com/en/user-guide/performance-query-exploring.

Long Table Scan

A long table scan occurs where most of the processing time is spent servicing remote disk I/O. This query references partition… information available only from the Account Usage Store; note that latency of up to 45 minutes applies. Entries are only inserted in snowflake.account_usage.query_history after the statement runs completely or is cancelled.

```
SELECT query_id,
       warehouse_name,
       warehouse_size,
       partitions_scanned / partitions_total AS partition_scan_ratio,
       partitions_scanned,
       partitions_total
FROM   snowflake.account_usage.query_history
WHERE  warehouse_name IS NOT NULL
AND    execution_status    = 'SUCCESS'
AND    bytes_scanned        > 0
AND    total_elapsed_time > 1000
AND    ( partitions_scanned / partitions_total ) > 0.5;
```

Guidance on resolving long table scan is provided in Chapter 3.

Spills to Disk and Out of Memory

Spills to disk are identified by examining the bytes_spilled… attributes. This query references bytes_spilled… information available from the Account Usage Store; note that latency of up to 45 minutes applies.

```
SELECT query_id,
       warehouse_name,
       warehouse_size,
       bytes_spilled_to_local_storage,
       bytes_spilled_to_remote_storage,
       bytes_sent_over_the_network
```

```
FROM    snowflake.account_usage.query_history
WHERE   warehouse_name IS NOT NULL
AND     bytes_spilled_to_local_storage  > 0
AND     bytes_spilled_to_remote_storage > 0;
```

Guidance on resolving spills to disk and OOM is provided in Chapter 3. You can find more information on spills to disk at https://community.snowflake.com/s/article/Performance-impact-from-local-and-remote-disk-spilling.

Snowflake Support

In the event all of the previous does not identify a root cause for the issues encountered, Snowflake Support is the first point of contact. During the writing of this book, I used trial accounts and found Snowflake Support very responsive and helpful. I recommend contacting Snowflake Support where required, even for trial accounts; you can find more information at https://www.snowflake.com/support/. You can raise a support case by following the guide at https://community.snowflake.com/s/article/How-To-Submit-a-Support-Case-in-Snowflake-Lodge.

Depending upon the level of support provided to your organization, there may be a sales engineer or performance expert dedicated to assisting you. The very best sales engineers proactively monitor consumption and will highlight problematic queries for your further investigation. I advise you to cultivate a good working relationship with your sales engineer and, where available, your performance expert too.

Snowflake Feature Use Cases

If you have arrived at this section without developing a deep understanding of your performance issue root cause, I recommend working through this chapter from the start. I do not recommend blindly implementing any feature without a full understanding of the potential impact.

In this chapter, you have examined many performance tuning tips to both reduce costs and reduce query runtimes. I expect these tips to deliver impactful business benefits, and I recommend both thorough investigation and testing before proceeding further.

In addition to the advice and guidance presented in this book, Snowflake presents several features aimed at remediating performance issues. Knowing when and how to enable Snowflake features is not just an important aspect of performance tuning; there is an implicit assumption that all code has been optimized before arriving at this section.

All features referenced in the following sections make use of serverless compute; see `https://docs.snowflake.com/en/user-guide/cost-understanding-compute#serverless-features`.

Let's now investigate these features.

Automatic Clustering

Automatic clustering works well for large tables; Snowflake recommends adding clustering keys for tables of 1TB or larger, but this is not a hard-and-fast rule. For optimal micro-partition pruning, clustering keys should match query predicates. Where date ranges are frequently selected, reducing the cardinality of dates to months by adding an extra attribute during data load may provide a performance benefit when predicates match.

Snowflake implements asynchronous automatic clustering via a background process that periodically reorganizes a small set of micro-partitions to achieve an acceptable performance standard.

Automatic clustering does not work well for query predicates that do not match the clustering keys. High-velocity, low-volume DML operations may overwhelm the automatic clustering service capability to re-cluster before the next batch of changes arrive.

Re-clustering incurs compute cost, and significant additional storage costs may accrue as micro-partitions are immutable.

Refer to Chapter 5 for a detailed investigation of automatic clustering.

Materialized Views

Currently, a materialized view can exist on a single table only. Materialized views work well for tightly focused subsets of data. Materialized views created for pre-aggregated or prefiltered data sets prevent expensive warehouse operations where the data is frequently accessed. Query plans may prefer a materialized view over base table access.

Materialized views should not be used where the contents are largely similar to the parent table. Materialized views should not be used for simple queries, i.e., those without aggregates or filters. High-velocity, low-volume DML operations may overwhelm the materialized view service capability to maintain materialized views before the next batch of changes arrive.

Maintaining materialized views incurs both compute cost and additional storage costs.

Search Optimization

Search optimization prebuilds optimized data structures called *search access paths* predicated upon high-cardinality attribute values spread across many micro-partitions. Search optimization prefers accessing very small subsets of data via equality predicates mapped via "search access paths." Snowflake recommends usage for tables of 100GB or larger; costs will be prohibitive for tables of less than 10GB. Search optimization may be implemented against tables with clustering keys where predicates do not match the clustering key or for unclustered tables.

Search optimization should not be used for accessing large sets of filtered data or for "search access paths" built on low-cardinality data sets. Inequality predicate matches are not suitable for search optimization. You can find more information on supported predicate types at `https://docs.snowflake.com/en/user-guide/search-optimization/queries-that-benefit#supported-predicate-types.`

High-velocity, low-volume DML operations may overwhelm the search optimization service capability to maintain "search access paths" before the next batch of changes arrive.

Maintaining search optimization incurs both compute cost and additional storage costs.

Refer to Chapter 7 for a detailed investigation of search optimization.

Query Acceleration

Query acceleration adds processing units to an existing warehouse as demand increases without increasing size. This is useful where the workload does not justify spinning up additional warehouse clusters, but the occasional availability of extra processing units would prevent queueing. The query optimizer may make use of extra processing units to parallelize some operations specifically for large table scans and ad hoc analytics. Mixed workloads may also benefit from query acceleration.

Before enabling query acceleration, test the existing configuration by decreasing the warehouse size and/or number of clusters to reduce the number of available processing units and then enable query acceleration. Look for queueing under the normal system load conditions.

Query acceleration incurs compute cost only; no additional storage costs accrue.

Refer to Chapter 6 for a detailed investigation of query acceleration.

Resource Monitors

Snowflake provides resource monitors as a means to control warehouse consumption, which is a reactive approach to limiting costs once the specified threshold has been reached.

Serverless Compute

Snowflake features increasingly offer serverless compute for cost-effective and simple implementation, but costs can quickly escalate.

The following table illustrates serverless compute components along with a brief summary of capabilities provisioned as derived from `https://docs.snowflake.com/en/user-guide/cost-understanding-compute#serverless-credit-usage`.

Component	Feature	Compute
Automatic Clustering	Automated background maintenance of each clustered table, including initial clustering and reclustering as needed	Serverless only
External Tables	Automated refreshing of the external table metadata with the latest set of associated files in the external stage and path	Serverless only
Materialized Views	Automated background synchronization of each materialized view with changes in the base table for the view	Serverless only
Query Acceleration Service	Execution of portions of eligible queries	Serverless only

(continued)

Component	Feature	Compute
Replication	Automated copying of data between accounts, including initial replication and maintenance as needed	Serverless only
Search Optimization Service	Automated background maintenance of the search access paths used by the search optimization service	Serverless only
Snowpipe	Automated processing of file loading requests for each pipe object	Serverless or warehouse
Snowpipe Streaming	Automated processing of file loading requests for each pipe object; currently INSERT only	Serverless only
Tasks	Scheduled tasks	Serverless or warehouse

Testing Code Changes

Most of the performance tuning advice provided in this chapter involves making invasive code changes. A recurring theme is to test once, test twice, and then test again. Make no apology for insisting upon full testing using production-like workloads. Executive management requires low-risk, high-value delivery, and your testing must reflect best practices at all times.

Summary

This chapter began by considering design decisions that have the most decisive impact on system performance. Tuning the design is the most important advice available *before* attempting to write any code and applies ubiquitously to all platforms, not just Snowflake.

The chapter next identified the available tools to aid your investigations and provided template code (several earlier chapters have deep dives and full explanations for the identified tooling).

Optimizing code, particularly where "lift and shift" from a legacy RDBMS has been performed, is invasive. Colleagues must be educated on Snowflake-specific performance requirements. Then the codebase must be refactored to take best advantage of Snowflake optimizer preferences and structures.

You then investigated how to identify object dependencies and constraints before examining how various objects are managed and differ from each other.

With a firm grasp of how to fix issues, you then investigated how to identify issues. Having the correct tools in hand along with sufficient context, you learned where to look and what to look for.

Your tuning approach should also be informed by the operating environment; nonproduction environments may be able to tolerate warehouses operating in economy mode, whereas production environments should implement warehouses operating in standard mode.

I discussed Snowflake features with the caution to consider them *after* all other code optimizations have been applied. Treat the root cause, not the symptoms!

Finally, I hope to have dispelled the myth of solely managing performance by resizing warehouses. Applying best practices goes a long way to both preventing performance issues from arising and facilitating reductions in warehouse size.

Afterword

Applying the knowledge gained from this book is a challenge I cannot prepare you for, as each scenario will present itself differently. However, you now have the toolkit in your hands to meet each challenge with confidence. Have confidence when investigating and remediating issues, and improve your technical real estate while saving both costs and time for your organization.

I conclude this book with the hope and expectation you have learned something new. In fact, I did not realize how little I knew when setting out to write this book; the journey has been enlightening to say the least.

With my very best wishes for your Snowflake journey, I look forward to seeing you at Snowflake Summit and various speaking engagements and hearing how this book made a difference.

Installing Python and the Tooling You Will Need

This appendix covers how to install the tooling referenced in Chapter 6.

Installing Python from the Command Line

Later in this appendix you will learn how to develop a Python parallel process to invoke several concurrent queries. In conjunction with Chapter 6, I show how to invoke parallel processes to simulate a load test as a starting point for stressing your applications to find out where they could break.

Load testing serves several purposes:

- Optimize your warehouse size for known workloads

- Identify spills, queueing, blocking, and out-of-memory scenarios

- Monitor trends to enable early intervention and prevent failure

In this appendix, I discuss how to install Python on your operating system; these instructions are generic in nature and work for many common desktops.

In conjunction with Chapter 6 the testing in this appendix proves there are version dependencies across both Snowflake-supported Python versions and tooling that may not easily be resolved. This appendix assumes Microsoft Windows 10.

© Andrew Carruthers 2024
A. Carruthers, *Tuning the Snowflake Data Cloud*, https://doi.org/10.1007/979-8-8688-0379-6

Checking the Installed Python Version

Snowflake does not support all versions of Python. At the time of writing, only versions 3.8, 3.9, and 3.10 are supported

To identify the currently installed Python version (if any), open a command window and type python --version, as shown in Figure A-1.

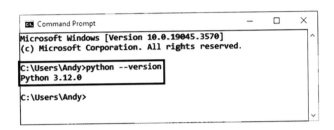

Figure A-1. *Checking the installed Python version*

As you can see in Figure A-1, this installed Python version is supported; therefore, you must downgrade to a lower version.

You can find further information on the supported Python versions at https://docs.snowflake.com/en/developer-guide/snowpark/python/setup#prerequisites.

Downgrading the Python Version

Downgrading Python involves invoking the Windows "Add or remove programs" feature by typing **add or remove programs** into the search bar, as shown in Figure A-2.

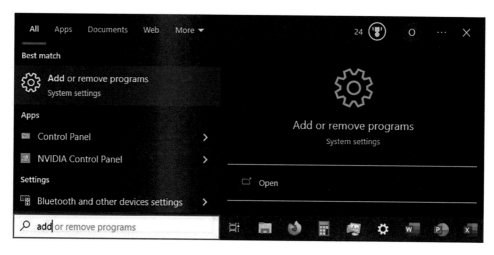

Figure A-2. *Adding or removing programs*

Within the search results, navigate to the version of Python to uninstall, in this example version 3.12, as shown in Figure A-3.

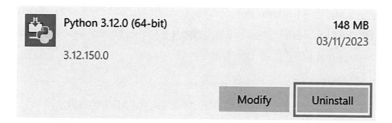

Figure A-3. *Uninstalling Python*

Click Uninstall after which a confirmation dialog will appear, as shown in Figure A-4. Click Close to finish.

Figure A-4. *Uninstall complete*

Installing Python

The Python installation process is dependent upon your operating system, Python availability, and Snowflake-supported version.

Before deciding on the Python version to install, please check the Snowflake prerequisites at `https://docs.snowflake.com/en/developer-guide/snowpark/python/setup#prerequisites`.

Installers for older versions of Python are periodically removed.

Once you have identified the Python compatibility, download the corresponding Python installer from `https://www.python.org/downloads/`. Alternatively, if looking for a specific Python version to support a known compatible configuration where installers have been removed from the main download location, these URLs may prove helpful:

- **Python 3.11:** `https://www.python.org/downloads/release/python-3110/`

- **Python 3.10:** `https://www.python.org/downloads/release/python-3100/`

- **Python 3.9:** `https://www.python.org/downloads/release/python-390/`

- **Python 3.8:** `https://www.python.org/downloads/release/python-380/`

I am using the Windows Installer (64-bit) noting the version recommendation.

At the time of writing, Python 3.11 is in public preview; I am therefore using Python 3.10 found here: `https://www.python.org/downloads/release/python-3100/`.

When invoking the installer, you should select the Install Now option and avoid using the custom installer. You may also be prompted to install the Python Launcher; leave this checkbox enabled.

Ensure the Add Python 3.10 to PATH checkbox is selected.

Then click Install Now. Assuming the setup is successful, click Close to complete the installation. Close any open command windows and re-open command window to pick up your latest installed Python version.

Installing Snowpark Python

You must also install Snowpark Python as a prerequisite for later creating a stand-alone executable. To do this, you can use `pip`, which should have been installed as part of the Python installation; if it wasn't, you can find further information at `https://pip.pypa.io/en/stable/installation/`.

Confirm the Python version by opening a command window and then typing the following:

```
python --version
```

Before proceeding, ensure your Python version meets the requirements.

With `pip` available, type this:

```
pip install snowflake-snowpark-python
```

Figure A-5 shows a successful installation.

```
Command Prompt
C:\Users\Andy>pip install snowflake-snowpark-python
Collecting snowflake-snowpark-python
   Downloading snowflake_snowpark_python-1.10.0-py3-none-any.whl.metadata (45 kB)
   -------------------------------------- 45.7/45.7 kB 753.5 kB/s eta 0:00:00

Downloading snowflake_snowpark_python-1.10.0-py3-none-any.whl (330 kB)
   -------------------------------------- 331.0/331.0 kB 1.7 MB/s eta 0:00:00
Installing collected packages: snowflake-snowpark-python
Successfully installed snowflake-snowpark-python-1.10.0
```

Figure A-5. *Snowflake-Snowpark-Python install complete*

You can find further information about snowflake-snowpark-python at `https://pypi.org/project/snowflake-snowpark-python/`.

Installing Pyinstaller and pip

Later in this appendix you will create a stand-alone Python executable for which you will use Pyinstaller.

You can install Pyinstaller by running the following:

```
pip install pyinstaller
```

`pip` is a package manager for Python packages.

When the Pyinstaller installation is complete, you may be prompted to upgrade `pip`. To do this, run the following:

```
python.exe -m pip install --upgrade pip
```

Figure A-6 shows the steps to take for both commands.

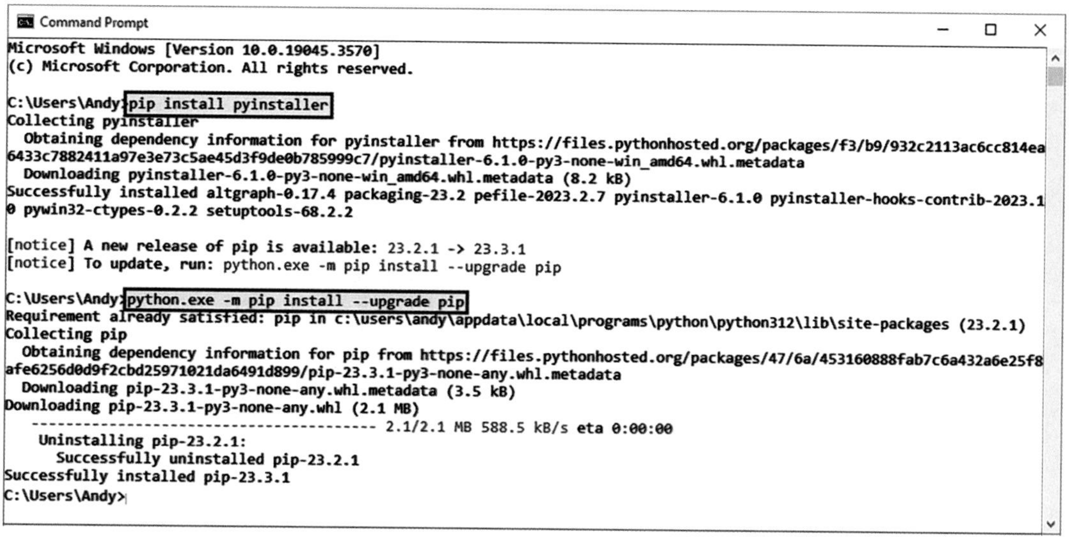

Figure A-6. *Installing Pyinstaller and pip*

You can find further information about pip at `https://www.w3schools.com/python/python_pip.asp`, and you can find Pyinstaller at `https://www.pyinstaller.org/en/stable/operating-mode.html`.

Installing Pandas and jinja2

Next, install Pandas.

```
pip install snowflake-snowpark-python[pandas]
```

Figure A-7 shows a successful installation.

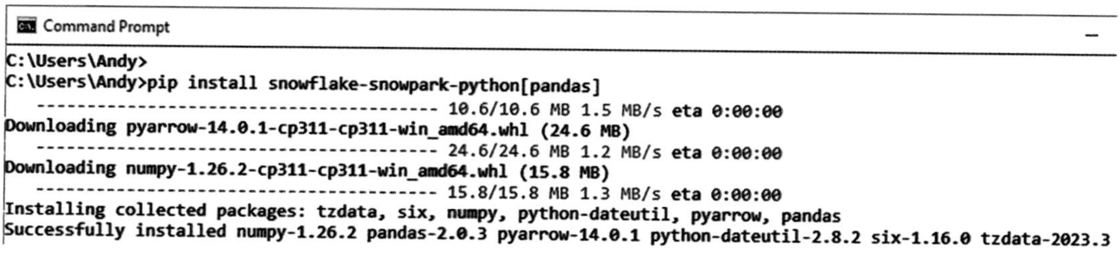

Figure A-7. *Installing Pandas*

You may also need to install jinja2.

```
pip install jinja2
```

Enabling Anaconda Packages

Preparing the environment requires the creation of Python stored procedures, for which you must use the ORGADMIN role. Attempting to use Python before accepting terms results in this error message:

> *"Anaconda terms must be accepted by ORGADMIN to use Anaconda 3rd party packages. Please follow the instructions at* `https://docs.snow-flake.com/en/developer-guide/udf/python/udf-python-packages.html#using-third-party-packages-from-anaconda.`*"*

Note the inclusion of a URL containing instructions for your reference: `https://docs.snowflake.com/en/developer-guide/udf/python/udf-python-packages.html#using-third-party-packages-from-anaconda`.

Figure A-8 shows the Snowsight-abbreviated navigation required to enable Anaconda packages.

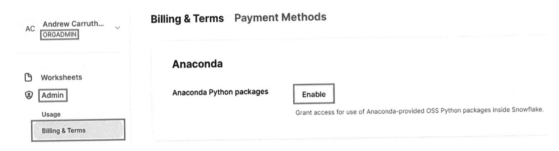

Figure A-8. *Enabling Anaconda Python*

Once enabled, a confirmation dialog appears, as shown in Figure A-9.

Figure A-9. *Anaconda package confirmation*

Anaconda package enablement takes a few minutes, and then an acknowledgment message is returned to the Snowsight console (not shown).

Downloading and Installing SnowCD

In the event you experience connectivity issues when attempting to use Python, you may find SnowCD helpful.

SnowCD is a Snowflake-supplied connectivity diagnostics tool. You can download it from `https://developers.snowflake.com/snowcd/`. I recommend downloading and installing SnowCD in the event you encounter connectivity issues.

After installing SnowCD, the dialog shown in Figure A-10 appears.

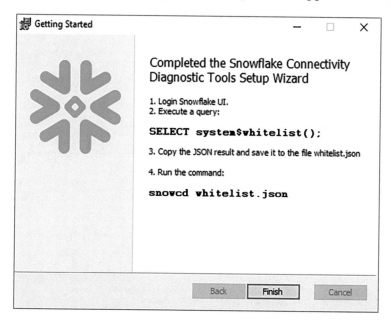

Figure A-10. *SnowCD completion dialog (on Windows)*

The SnowCD completion dialog at the time of writing has not been updated; `system$whitelist` has been deprecated, and from within the Snowflake user interface the correct command to execute is as follows:

```
USE ROLE accountadmin;

SELECT system$allowlist();
```

Then follow these steps:

1. Copy the resultant JSON record into memory.

2. Create a new file called `whitelist.txt` in the same directory as your command window, in this example, `C:\Users\Andy` (your default directory will differ).

3. Paste the copied JSON into `whitelist.txt`.

4. Close the file and rename it to `whitelist.json`.

5. Run `snowcd whitelist.json`.

Figure A-11 shows the expected response: "All checks passed." Troubleshooting is beyond the scope of this chapter, so in the event that you see an error, please refer to the documentation at `https://docs.snowflake.com/en/user-guide/snowcd`.

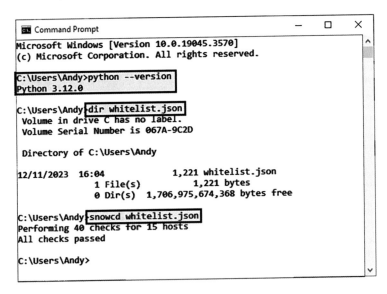

Figure A-11. *SnowCD check completion*

Index

Printed in the United States
by Baker & Taylor Publisher Services